W0017525

Budget Estimating Relationships for Depot-Level Reparables in the Air Force Flying Hour Program

Gregory G. Hildebrandt

Prepared for the United States Air Force

PROJECT AIR FORCE

The research described in this report was sponsored by the United States Air Force under Contracts F49642-01-C-0003 and FA7014-06-C-0001. Further information may be obtained from the Strategic Planning Division, Directorate of Plans, Hq USAF.

Library of Congress Cataloging-in-Publication Data

Hildebrandt, Gregory G.
    Budget estimating relationships for depot-level reparables in the Air Force flying hour program / Gregory G. Hildebrandt.
        p. cm.
    Includes bibliographical references.
    ISBN 978-0-8330-4121-0 (pbk. : alk. paper)
    1. Airplanes, Military—United States—Parts—Costs. 2. United States. Air Force—Equipment—Maintenance and repair—Costs. 3. United States. Air Force—Accounting. I. Project Air Force (U.S.) II. Title.

UG1243.H55 2007
358.4'18—dc22

2007023609

The RAND Corporation is a nonprofit research organization providing objective analysis and effective solutions that address the challenges facing the public and private sectors around the world. RAND's publications do not necessarily reflect the opinions of its research clients and sponsors.

RAND® is a registered trademark.

Published 2007 by the RAND Corporation
1776 Main Street, P.O. Box 2138, Santa Monica, CA 90407-2138
1200 South Hayes Street, Arlington, VA 22202-5050
4570 Fifth Avenue, Suite 600, Pittsburgh, PA 15213-2665
RAND URL: http://www.rand.org/
To order RAND documents or to obtain additional information, contact Distribution Services: Telephone: (310) 451-7002;
Fax: (310) 451-6915; Email: order@rand.org

Preface

Cost estimating relationships (CERs) and budget estimating relationships (BERs) are empirical relationships depicting how key variables affect relevant costs and budgets. These relationships are extremely useful in many types of defense analyses, including long-range planning, the analysis of force structure alternatives, cost-effectiveness analysis supporting Analysis of Alternatives studies, cost estimation during the acquisition process, and support provided to the budgetary process. This document focuses on BERs, which explain the obligation of Operations and Maintenance (O&M) funds to satisfy the demand, by maintenance organizations of the United States Air Force (USAF) major commands (MAJCOMs), for flying depot-level reparables (DLRs).

Currently, Operating and Support (O&S) CERs and BERs are limited, and USAF analysts must frequently develop detailed aircraft O&S estimates for situations in which estimating relationships are more appropriate. The principal objectives of this task are to improve the tools available by developing relationships for various O&S categories. During this phase of the project, the focus has been on developing BERs, which explain the direct effect of specified variables on the obligations associated with the reparable parts. Flying DLRs constitute this budget category. CERs and BERs are also being developed for aircraft and engine overhauls, base-level maintenance, and aircraft modifications. These are O&S categories for which costs and budgets are interrelated to those for DLRs.

This research was part of a multiyear project entitled "Weapon System Costing," sponsored by SAF/AQ and conducted within the Resource Management Program of RAND Project AIR FORCE. Each year, a number of research tasks are selected by the sponsor for inclusion in the umbrella project. The purpose of the project is to develop better cost estimating methods for use by the acquisition community, examine the impacts of USAF and Department of Defense policies on weapon system costs, and establish a Center for Excellence in Cost Analysis at the RAND Corporation.

The research reported here was sponsored by Lt Gen Donald J. Hoffman, Principal Deputy, Office of the Assistant Secretary of the Air Force, Acquisition (SAF/AQ), and Blaise J. Durante, Office of the Assistant Secretary of the Air Force, Acquisition Integration (SAF/AQX). The study's technical monitor was Jay Jordan, technical director of the Air Force Cost Analysis Agency.

Other RAND Project AIR FORCE documents that address weapon system acquisition and cost estimating issues include the following:

- *An Overview of Acquisition Reform Cost Savings Estimates* (MR-1329-AF), by Mark A. Lorell and John C. Graser.
- *Military Airframe Acquisition Costs: The Effects of Lean Manufacturing* (MR-1325-AF), by Cynthia R. Cook and John C. Graser.
- *Military Airframe Costs: The Effects of Advanced Materials and Manufacturing Processes* (MR-1370-AF), by Obaid Younossi, Michael Kennedy, and John C. Graser.
- *Military Jet Engine Acquisition: Technology Basics and Cost-Estimating Methodology* (MR-1596-AF), by Obaid Younossi, Mark V. Arena, Richard M. Moore, Mark A. Lorell, Joanna Mason, and John C. Graser.
- *Test and Evaluation Trends and Costs in Aircraft and Guided Weapons* (MG-109-AF), by Bernard Fox, Michael Boito, John C. Graser, and Obaid Younossi.
- *Software Cost Estimation and Sizing Methods: Issues and Guidelines* (MG-269-AF), by Shari Lawrence Pfleeger, Felicia Wu, and Rosalind Lewis.

- *Lessons Learned from the F/A-22 and F/A-18E/F Development Programs* (MG-276-AF), by Obaid Younossi, David E. Stem, Mark A. Lorell, and Frances M. Lussier.
- *Price-Based Acquisition: Issues and Challenges for Defense Department Procurement of Weapon Systems* (MG-337-AF), by Mark A. Lorell, John C. Graser, and Cynthia R. Cook.
- *Impossible Certainty: Cost Risk Analysis for Air Force Systems* (MG-415-AF), by Mark V. Arena, Obaid Younossi, Lionel A. Galway, Bernard Fox, John C. Graser, Jerry M. Sollinger, Felicia Wu, and Carolyn Wong.
- *Systems Engineering and Program Management: Trends and Costs for Aircraft and Guided Weapons Programs* (MG-413-AF), by David E. Stem, Michael Boito, and Obaid Younossi.
- *Evolutionary Acquisition: Implementation Challenges for Defense Space Programs* (MG-431-AF), by Mark A. Lorell, Julia F. Lowell, and Obaid Younossi.
- *Historical Cost Growth of Completed Weapon System Programs* (TR-343-AF), by Mark V. Arena, Robert S. Leonard, Sheila E. Murray, and Obaid Younossi.

In addition to members of the defense acquisition community, these reports are of interest to those in the national security community who are involved in analyzing alternative military postures, and to members of the aircraft industry's analytical community.

RAND Project AIR FORCE

RAND Project AIR FORCE (PAF), a division of the RAND Corporation, is the U.S. Air Force's federally funded research and development center for studies and analyses. PAF provides the Air Force with independent analyses of policy alternatives affecting the development, employment, combat readiness, and support of current and future aerospace forces. Research is conducted in four programs: Aerospace Force Development; Manpower, Personnel, and Training; Resource Management; and Strategy and Doctrine.

Additional information about PAF is available on our Web site at http://www.rand.org/paf/.

Contents

Figures

Tables

Summary

One of the key cost and budget O&S categories is flying DLRs. These are the reparable spare parts that directly support the USAF Flying Hour Program. In fiscal year 2002 (FY02), they constituted about $3.5 billion of total O&S costs of about $24.3 billion, or slightly greater than 14 percent of the total. It can be argued, however, that it is a particularly important 14 percent. Because of the direct connection between flying hours and military readiness, fulfilling the Flying Hour Program is one of the priority objectives of the Air Force. As a result, the funds budgeted for this program, within a MAJCOM, are fenced and cannot be moved to other Air Force budget activities without Chief of Staff approval. And, if the budgeted funds are not adequate to support this program, and additional funding is not authorized by Congress, funds needed to support the program may be moved from certain O&M activities by the MAJCOM. As the scale of these other activities is reduced, their contribution to military readiness would decline.

The O&M-obligated funds finance the "net sales" of the spare parts by the Materiel Support Division (MSD) of Air Force Materiel Command (AFMC) to the commands and their parts-demanding organizations. While other terms have been used, throughout this analysis, we use the term "net sales" to describe the dollar value of the flying DLR transactions. Because the associated obligations are sanctioned by budget authorizations, the models that explain net sales are called budget estimating relationships, or BERs.

Objective of the Study

The purpose of this research is to support the Air Force Cost Analysis Improvement Group (AFCAIG) flying DLRs budgeting process by explaining why net sales of these DLRs to the commands are at their historic levels. The AFCAIG analyzes command inputs on numerous aircraft Mission Design Series (MDS), submits budget recommendations to the Air Force corporate structure, and develops command-specific cost per flying hour (CPFH) factors. To both explain the historical data and provide the AFCAIG with a tool to better understand the commands' budgetary submissions, we develop several explanatory BERs to understand why flying DLRs are at their particular levels.

We explain the historical flying DLRs by estimating models that relate net sales to the contemporaneous values of aircraft characteristics, operational tempo (OPTEMPO), and time-related variables. The aircraft characteristics are aircraft mission type and flyaway cost; the OPTEMPO variables are flying hours, average sortie duration (ASD), and landings per sortie, which capture the macro usage dimensions of mission profiles; and the time-related variables are fiscal year categorical variables and aircraft MDS age. We also extend the analysis to account for the serial correlation that occurs across fiscal years.

Explanatory Variables in the BER

If we focus our attention on flying DLR transactions during a particular period, and successfully explain how these budgets change when specified variables change, the causal structure of flying DLR budget determination is identified. To achieve this objective, we develop BER response schedules using data for explanatory variables that are contemporaneous with flying DLR net sales.[1] Such response schedules

---

[1] Causality and the role played by response schedules are discussed in Freedman (2005). Simon (1990) discusses causality in terms of specified mechanism by "how widely we draw the boundaries of the system to be examined." The boundaries of this analysis exclude intertemporal phenomenon. We are also isolating the transactional relationship between the base maintenance organization and base supply from other elements of the supply chain and repair cycle.

permit one to estimate how DLR budgets respond to changes in the explanatory variables. As a result, for an aircraft MDS, if a change in flying hours is instituted, or there is a change in the retirement profile that affects the average age of the remaining aircraft, the model can estimate the effect of such a policy intervention.

Intertemporal Prediction

It is likely that net sales of flying DLRs are also influenced by factors occurring in prior periods. While it is quite difficult to develop an explanatory model that characterized the full nature of intertemporal effects, we can capture the broad forces associated with intertemporal association by exploiting the serial correlation that exists among the residuals of subsequent periods. We do this and find that the results of the contemporaneous causality model are robust. This second model is particularly useful for predicting net sales.

Statistical Methods Employed

Because of the limited data for each of the aircraft MDS, we employ longitudinal regression statistical methods and explain flying DLR budgets by analyzing all of the aircraft MDS-command data simultaneously.

We must emphasize that there are significant correlations among the explanatory variables, but that the multiple-regression estimation techniques employed are designed to take these into account. Whenever we speak of an explanatory variable having a positive or negative effect on net sales, we always mean that the values of the other explanatory variables in the model are formally held constant. Our shorthand for this is "other things equal."

Empirical Findings

The first model hypothesized includes fiscal year categorical variables, and we find that these are not statistically significant and are deleted from the analysis (see p. 42).

In the revised model, which excludes the fiscal year variables, all aircraft mission types show significantly lower net sales than fighters, "other things equal." We also show a significant relationship between flyaway cost and net sales. A 1 percent increase in flyaway cost increases net sales by about 0.81 percent (see p. 44).

In this aircraft system model, a 1-percent increase in flying hours increases net sales by about 1.04 percent. The flying hours effect is so close to a proportional relationship between flying hours and net sales that it supports the working assumption of the CPFH AFCAIG budgeting process that there is a proportional relationship between net sales and flying hours (see p. 44). Under this assumption, changes in flying hours result in like-proportional changes in net sales.

During peacetime training activities, there have traditionally been very high correlations among flying hours, sorties, and landings, which are three different measures of OPTEMPO. As a result, in this peacetime training environment, it is very difficult to separate the different effects of these three variables on net sales. However, during contingencies, mission profiles change. ASD increases (see p. 31) and landings per sortie decrease (see p. 32). Because of the significant amount of contingency flying in FY02 and FY03, there is now sufficient independent variation in ASD and landings per sortie to incorporate these variables in BERs. We find that ASD has a negative, statistically significant effect on net sales, while landings per sortie have a significant positive effect.

Aircraft Aging

For the past several years, there has been a strong interest in the aging effects. When the traditional ordinary least squares (OLS) estimation technique is used, we show that a one-year increase in MDS age increases net sales by about 2.7 percent (see p. 45). However, we emphasize that the measured aging effect must be given a broader interpretation than the simple effect of material degradation. It can also embody the effect of aircraft modifications, and may be affected by technical progress that occurs over time, as new aircraft MDS enter the inventory.

Observations on BERs

Our discussion of the explanatory models notes many of the complications associated with constructing such models. Yet in some settings—say, during the acquisition process—the values of contemporaneous variables provide the primary input to estimating cost. In such analyses, aircraft type, OPTEMPO, and aircraft age variables are very natural variables to consider. To aid in the validation of budget estimates developed by the MAJCOMs, and to understand the moving forces determining net sales, these models may play a useful role. Also, to predict the value of flying DLRs, the model that accounts for the correlation among the residuals has attractive properties.

The appendix includes several amplifications and extensions. We address the issue of outliers, and investigate a first-difference model to better isolate the pure effects of aging and eliminate any possible "non-stationarity" from the data.

Acknowledgments

Dr. Bob Roll, director of RAND Project AIR FORCE's Resource Management Program, provided outstanding intellectual leadership throughout this project. Working with members of the Air Force acquisition and cost community, Bob organized the project and identified the primary lines of inquiry. Drawing on his extensive background in logistics, cost analysis, and economics, Bob provided frequent advice on both statistical and econometric methodology. He emphasized the complementary relationship between the proper use of principles from accounting, including an in-depth understanding of the meaning of the variables of interest, and the statistical and econometric methods employed. He also understood the importance of a thorough interdisciplinary review as an integral part of a report's preparation. Carol Fan, Lionel Galway, Pierre-Carl Michaud, and Raymond Pyles provided insightful comments that contributed significantly to the quality of this analysis. Bob passed away recently, and I hope that the final version of this report sustains his passion for objective analysis of the highest quality being conducted by RAND. Bob Roll possessed skills in numerous areas of defense policy that extend beyond cost and logistics analysis, and was a national resource.

Abbreviations

ACC	Air Combat Command
AFCAIG	Air Force Cost Analysis Improvement Group
AFMC	Air Force Materiel Command
AFTOC	Air Force Total Ownership Cost
AFWCF	Air Force Working Capital Fund
AMC	Air Mobility Command
ANG	Air National Guard
ASD	average sortie duration
BER	budget estimating relationship
CAIG	Cost Analysis Improvement Group
CER	cost estimating relationship
CLS	contractor logistics support
CPFH	cost per flying hour
DLR	depot-level reparable
DMAG	Depot Maintenance Activity Group
EEIC	Element of Expense/Investment Code
ERRC	expendability/recoverability/reparability/cost

FSC	Federal Supply Class
FY	fiscal year
GLS	generalized least squares
GSD	General Support Division
IMPAC	International Merchant Purchase Authorization Card
IOD	initial operational delivery
ISR	intelligence, surveillance, and reconnaissance
LDT	Logistics Distribution Table
MAJCOM	major command
MDS	Mission Design Series
MSD	Materiel Support Division
NIIN	National Item Identification Number
NRTS	Not Reparable This Station
NSN	National Stock Number
O&M	Operations and Maintenance
O&S	Operating and Support
OLS	ordinary least squares
OPTEMPO	operational tempo
ORG	organization code
PAA	primary aircraft authorized
PAF	RAND Project AIR FORCE
PDS	Program Data System
PEC	Program Element Code

Q	quarter
REMIS	Reliability and Maintainability Information System
SAF/FM	Secretary of the Air Force, Financial Management
SAF/FMC	Secretary of the Air Force, Financial Management (Cost and Economic Analysis)
SBSS	Standard Base Supply System
SEE	standard error of the estimate
SMAG	Supply Management Activity Group
SRAN	Stock Record Account Number
TAI	total aircraft (or active) inventory
TOC	Type Organizational Code
USAF	United States Air Force

CHAPTER ONE

Introduction

The primary purpose of this analysis is to develop budget estimating relationships (BERs) for flying depot-level reparables (DLRs). It is expected that these relationships will support the flying program's programming and budgeting process for those spare parts purchased by the major commands (MAJCOMs). These purchases are made from the Materiel Support Division (MSD) of Air Force Materiel Command (AFMC). In addition to operating the supply system and conducting component repair (and overhaul) at the depots, MSD supports the Air Force Working Capital Fund (AFWCF) by setting administrative prices that apply to the transactions associated with the spare parts.

In this chapter, we provide some background on the spare parts transactions between the MAJCOMs and MSD. This includes a discussion of DLRs and a subgroup of this category, called "flying DLRs." Chapter Two provides a discussion of the data employed in the analysis; Chapter Three addresses the specification of the models estimated; Chapter Four contains the empirical results; and Chapter Five contains the primary conclusions of the analysis.[1]

---

[1] Other RAND work that relates to this study can be found in Pyles (2003). Pyles develops several cost estimating relationships (CERs), including one for the DLRs identified in the AFMC D200 database. These DLRs, which are valued at latest repair cost, are projections of the demands for component repair at the depot. In this study, we focus on the net sales of DLRs, valued at established prices, at the time the funds are obligated. Hildebrandt and Sze (1990) use statistical methods to estimate Operating and Support (O&S) costs for a range of categories. While some of the explanatory variables selected are similar to those used in this study, the earlier research does not address DLRs, which, as the term is used now, did not

It is important to note that "DLRs" is now a colloquial expression. DLRs include both MSD-managed reparables and consumables. However, what are called DLRs in MSD includes both parts that are reparable at the depot and certain consumables (parts which are typically not repaired and are disposed of after use). The AFWCF MSD was formed in fiscal year 1998 (FY98) by combining the Reparable Support Division and the System Support Division, where the former contained primarily depot reparables and the latter contained base-level consumables unique to a particular system. However, more than 95 percent of the DLR transactions are associated with depot reparables, and this may explain the continued use of the term DLRs.

A review of the transactions process using demand and supply as organizing categories is shown in Figure 1.1. Command base-level maintenance organizations demand DLRs held by base supply, from MSD's Supply Management Activity Group (SMAG), to replace "broken" parts that are turned in. Through the obligation of Operations and Maintenance (O&M) funds, these parts are "purchased" by the commands from the SMAG and recorded in the Standard Base Supply System (SBSS) reporting system as a sale. For each individual part, an established price, set by MSD, is paid.

When the part turned in is designated as Not Reparable This Station (NRTS), there is no credit received for the part, and it is sent by the SMAG to the depot for either replacement or component repair by the Depot Maintenance Activity Group (DMAG). The obligations by the commands to purchase the replacement parts provide the funds required for parts acquisition and repair at the depot. The SMAG and DMAG are responsible for parts repair and acquisition as well as associated inventory management, and constitute the supply side of the equation.

Not indicated in the diagram, but an important element of the process, are those parts repaired in base-maintenance "backshops." When a part repaired in these shops is returned to base supply, SBSS, under SMAG accounting procedures, gives a credit to the parts-demanding

---

exist at the time of the study. An analysis of Army M1 data has been conducted by Peltz et al. (2004). They show that age is a significant predictor of part failures.

Figure 1.1
Cost Per Flying Hour (CPFH) Demand and Logistics Supply

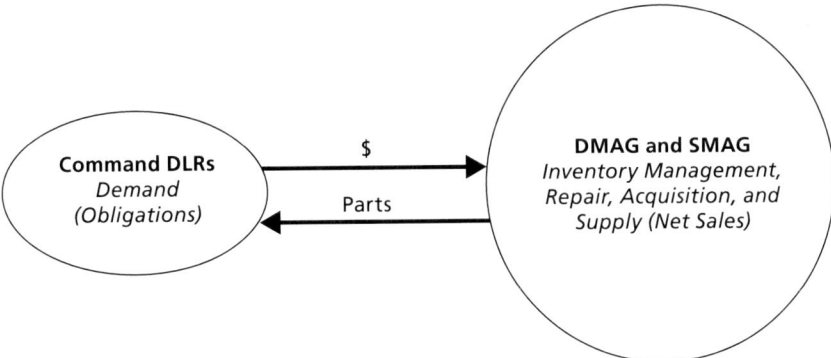

organization equal to the established price. When one aggregates across the relevant transactions, which are the sales less credits, the resulting net sales constitute the funds obligated during the period. We use "net sales" throughout this analysis to represent the dollar size of the transactional information.[2]

If a part is not available from base supply, it is back-ordered. Then, if such a part is received during the same fiscal year, the established price in effect that year applies to the transaction. However, should the part be delivered the following fiscal year, funds are obligated based on the next fiscal year's price, which is published toward the end of the fiscal year in which the part is back-ordered. These types of adjustments, however, are not captured by SBSS.

An obligation expense is incurred by the commands to acquire the parts recorded as net sales by MSD. From the standpoint of the commands, this expense, within a particular budget activity, represents a cost because alternative flying program activities within a particular budget activity are being foregone. O&M funds are obligated that have alternative uses within a MAJCOM Flying Hour Program for this budget activity. However, although these obligations are based

---

[2] If a maintenance organization holds a part for backshop repair, there could be a fiscal year timing disconnect between the net sales of reparables and parts designated NRTS.

on forecasted costs, at the time they occur they are not a direct cost to the Air Force. These obligations underwrite the cost authority needed to accomplish the repair and replenishment activities conducted by the depots. From the standpoint of the Air Force, the resource costs actually occur at the depot where the costs are incurred. Because of this distinction, we use the term "net sales" throughout the analysis. We recognize, however, that "cost" is frequently used, particularly in the context of CPFH calculations and discussions. Strictly speaking, however, what is a cost from the perspective of the commands is not necessarily a cost from the standpoint of the Air Force (or society).[3]

There are several databases maintained by the Air Force Total Ownership Cost (AFTOC) accounting system that contain data relevant to this analysis. One of these is called AFTOC-CAIG Format, where CAIG stands for the Cost Analysis Improvement Group, thereby indicating that the O&S accounting framework established by the Office of the Secretary of Defense, Program Analysis and Evaluation, and applicable to all the services, is being used.

For Air Force O&M expenditures, AFTOC-CAIG is based on information in the Command On-Line Accounting and Reporting System. Most of the DLR data contained the AFTOC-CAIG represent the SBSS-recorded transactions.

Definitions of Flying DLRs

It is important to understand that not all aircraft DLRs are designated as "flying DLRs," and that, in addition, there are two working definitions of flying DLRs. This distinction is clarified as we continue to describe the various databases.

The AFTOC-Commodities database (and also AFTOC-CAIG) summarizes both the reparables and consumables transactions reported in SBSS. In AFTOC-Commodities, "flying DLRs" are defined as those

---

[3] Other terms that have the same meaning as "cost" are "obligations," "obs," "demands," and "consumption." Because many parts are repaired and not consumed, consumption should be interpreted as the consumption of the services provided by the parts.

net sales from MSD to the commands under Element of Expense/ Investment Code (EEIC) 644. The EEIC 644 designation is assigned based on an MSD Budget Code and Type Organizational Code (TOC) designation.[4] Under this definition, organizations consuming supplies to perform maintenance on aircraft or tactical missile weapon systems and related support equipment have their transactions reported in EEIC 644. In contrast, EEIC 645 is assigned to DLRs that do not meet this definition. Included in EEIC 645 DLRs are parts for ground systems (e.g., radars).

The commands, when supporting the Air Force Cost Analysis Improvement Group (AFCAIG) process, employ a somewhat different definition of flying DLRs. Specific aircraft Mission Design Series (MDS)–designated organization codes (ORGs) acquiring parts from base supply (Stock Record Account Numbers, or SRANs) are coded as "fly" by the commands, and are included in the flying program analysis conducted by the commands. The information is recorded in AFTOC's Logistics Distribution Table (LDT). While these two definitions may sound the same, the first, using the Budget Code and TOC, is based on the Air Force's internal financial system, while the second identifies the ORGs associated with each specific aircraft MDS purchasing parts under the Flying Hour Program. There is, therefore, a strong logistical emphasis in the second definition. In practice, the flying DLR totals under the second definition are slightly smaller than under the first. In this analysis, the fly-coded data, provided by SAF/FMC, are employed.

Another data source, which we call "command generated," constitutes the data submitted to SAF/FMC during the AFCAIG process. It is understood that the command-generated data are based on the fly indicator business rules, include transactions not reported in SBSS, and contain adjustments for the repricing of parts on back order near the end of the fiscal year. At that time, next year's price schedule has been published and the commands' current-year back-order obligations are

---

4 MSD (Budget Code 8) plus a TOC 6, 7, 8, or 9 constitute EEIC 644. EEIC 645 would be defined by MSD plus a TOC not equal to 6, 7, 8, or 9. There is a discussion of this, as it relates to the data employed in this study, in AFTOC-Battelle (2004).

revised using these prices. Also, the commands can use local knowledge of the transactions to make necessary journal voucher adjustments to the information in the Air Force reporting systems.

The command-generated data submitted by the commands are reviewed by SAF/FMC and the Air Force corporate structure, and used to develop CPFH factors. These factors are based on the assumption that, over some relevant range, costs are proportional to flying hours and, therefore, can be used to predict demand for parts for a specified Flying Hour Program when there are marginal changes in flying hours.

Table 1.1 contains the CPFH factors for Budget Year 01 for selected fighters. In this analysis we are focusing on the data called

Table 1.1
Selected Fighter Flying Hour and Primary Aircraft Authorized (PAA) Factors

| | Budget Year 01 (FY01 Constant $) | | | | | | | |
| | Per Flying Hour Costs | | | | Per PAA Costs | | | |
MDS	Consumer Support GSD	DLR MSD	Aviation Fuel	Depot Maint.	Total Flying Hour Costs	Depot Maint.	Support Equipment	Total PAA Costs
F-15A	421	3,873	1,712	54	6,060	229,721	57,103	286,824
F-15B	421	3,842	1,712	53	6,028	254,446	57,103	311,549
F-15C	551	4,951	1,610	67	7,179	230,738	57,103	287,840
F-15D	530	4,935	1,613	69	7,147	206,080	57,103	263,183
F-15E	620	5,175	1,923	121	7,839	145,398	57,103	202,501
F-16A	309	2,293	867	103	3,572	23,130	41,075	64,205
F-16B	309	2,293	867	122	3,591	6,558	41,075	47,633
F-16C	332	2,472	917	54	3,775	35,059	41,075	76,134
F-16D	337	2,490	910	56	3,793	29,336	41,075	70,411

NOTE: The data portrayed in this table are based upon the results of the December 2000 AFCAIG-approved factors contained in AFI 65-503. The factors created include General Support Division, or GSD (this includes flying hour International Merchant Purchase Authorization Card, or IMPAC, purchases), DLRs, and aviation fuel. AFWCF has now ostensibly combined what was the System Support Division and DLR into a new commodity called the Materiel Support Division (MSD). The depot mainte-nance numbers only include organic maintenance costs and do not include CLS maintenance costs. Depot maintenance costs represent an average of three years (1997–99). The support equipment costs were inflated from last year's costs and will change as updates become available.

DLR MSD. Because these are described as "Per Flying Hour Costs," they implicitly represent only the flying DLRs.

Now that the various data systems have been reviewed, we can examine flying DLRs relative to other O&S categories to obtain a sense of its relative size. AFTOC-CAIG is the data source that covers all the O&S cost categories, and Table 1.2 shows that, in FY02, flying DLRs constitute about $3.5 billion of total O&S costs of $24.3 billion, or somewhat greater than 14 percent of the total.

Table 1.2
FY02 Total Air Force–Wide Aircraft O&S Costs, AFTOC-CAIG (millions of then-year dollars)

CAIG Level 1	CAIG Level 2	CAIG Level 3	TY $M
1.0 Mission Personnel	1.1 Operations		2,223.2
	1.2 Maintenance Off Equipment (1.2.2 - 1.2.4)	1.2.1 Organizational (On Equipment)	1,698.5
		1.2.2 Intermediate	1,362.1
		1.2.3 Ordnance Maintenance	577.5
		1.2.4 Other Maintenance Personnel	1,004.7
	1.2 Maintenance Total		4,642.8
	1.3 Other Mission Personnel		987.9
1.0 Mission Personnel Total			7,853.9
2.0 Unit-Level Consumption	2.1 Petroleum, Oil, and Lubricants (POL)/Energy Consumption		2,638.0
	2.2 Consumables	2.2.1 General Support Division	886.7
		2.2.2 System Support Division	0.3
		2.2.3 Mission Support Supplies	307.6
	2.2 Consumables Total		1,194.6
	2.3 DLRs	**2.3.1 Flying DLRs**	**3,472.5**
		2.3.2 Non-Flying DLRs	18.1
	2.3 DLRs Total		3,490.6
	2.4 Training Munitions		41.5
	2.5 Other Unit-Level Consumption		404.8
2.0 Unit-Level Consumption Total			7,769.4
3.0 Intermediate Maintenance			0.7
4.0 Depot Maintenance (not DLRs)	4.1 Aircraft Overhaul		1,530.4
	4.3 Engine Overhaul		669.7
	4.4 Other		102.1
4.0 Depot Maintenance (not DLRs) Total			2,302.2
5.0 Contractor Support			1,891.2
6.0 Sustaining Support			770.1
7.0 Indirect Support	7.1 Personnel Support		1,280.9
	7.2 Installation Support		2,476.0
7.0 Indirect Support Total			3,757.0
Total O&S Cost			24,344.4

It may be argued, however, that this is a particularly important 14 percent. Because of the direct connection between flying hours and military readiness, fulfilling the Flying Hour Program is one of the priority objectives of the Air Force. The funds in the Flying Hour Program are "fenced" by the Air Force, and Chief of Staff of the Air Force approval is needed to spend the funds elsewhere. If there is an expected overrun of the Flying Hour Program, and additional funding is not authorized by Congress, it is extremely likely that the funds needed to support the program will be drawn from other budgeted uses of these funds within the MAJCOM that are not subject to their own restrictions.[5]

---

[5]   The Flying Hour Program consists of flying DLRs (EEIC 644), General Support Division consumables (EEIC 609), IMPACs (EEIC 61952), and aviation fuel (EEIC 699). These are elements included in AFI 65-503, Table A2-1 (see Table 1.1 on page 6).

CHAPTER TWO

Description of Flying DLRs and Related Data Elements

In this chapter, the data used in the analysis are discussed. Particular attention is directed to the source of, and summary measures of, the flying DLR data. We also provide information on the other data elements used in the study.

Aircraft MDS Combinations

It is important to first note that during the AFCAIG process, SAF/FMC employs combinations of aircraft MDS in their analysis. Aircraft MDS are combined when one or more of the following conditions hold:

- Several aircraft MDS within a command share the same Program Element Code (PEC) and it is, therefore, difficult to break out the funds at a finer-grain level.
- There is an associated concern with maintaining consistency across commands.
- There is significant parts commonality across certain aircraft MDS.
- There is a low total aircraft (or active) inventory (TAI) level for aircraft MDS assigned to one or more of the commands.[1]

---

[1] The aircraft MDS is identified when the SRAN-ORG tables are developed by the commands. Primarily MDS-CAIG business rules are used to associate the data with aircraft MDS.

To enhance comparability with the AFCAIG process, we have used the same combinations of aircraft MDS. These aircraft MDS combinations and the associated predominant missions are contained in Table 2.1. For these aircraft, flyaway cost data were obtained from SAF/FMC.

Full contractor logistics support (CLS) aircraft are not included in the analysis. The flying DLR obligations are dependent on variegated factors specific to the individual CLS contract.[2] Also, for the remainder of this analysis, to avoid new terminology, these combinations are referred to as aircraft MDS.

Flying DLR Net Sales

The flying DLR data source employed in the explanatory models is the fly indicator data from the LDT that we have received from SAF/FMC. We obtained quarterly net sales for aircraft MDS-command-National Item Identification Number (NIIN)–specific transactions for the periods FY98Q1–FY03Q4.[3] The resulting database is quite large, and there are more than one million MDS-command-NIIN-fiscal year-quarter data points.

It may be helpful to understand the nature of the underlying transactions data used in the analysis. Table 2.2 provides some data for the F-16C/D in FY03Q4. The quarterly transactions data are provided for those NIINs with the 25 largest net sales during the quarter from a total of almost 7,000 F-16C/D-command-FY03Q4–identified NIINs with transactions. The table contains information for command, FSC,

---

[2] Full CLS aircraft have been identified by the CLS Integrated Product Team, and have been identified in Lively (2004). One full CLS aircraft we considered retaining is the Joint Stars E-8C. However, we included this aircraft in several modeling excursions, and the results changed very slightly. This aircraft MDS is not included in any of the empirical models discussed below.

[3] The NIIN of a particular spare part is a way of identifying it. The NIIN can be contrasted with the National Stock Number (NSN) of the part, which includes the three-digit Federal Supply Class (FSC). Because FSCs are not believed to be consistent over time, analysts frequently use the NIIN, which is the NSN less the FSC.

Table 2.1
Aircraft MDS Combinations and Associated Missions

Aircraft MDS Combination	Mission
A/OA-10A	Fighter
AC-130H	Special operations
AC-130U	Special operations
AT-38B	Trainer
B-1B	Bomber
B-2A	Bomber
B-52H	Bomber
C-130E/H	Cargo
C-130J	Cargo
C-135B/C/E	Cargo
C-141B/C	Cargo
C-5A/B/C	Cargo
E-3B/C	ISR[a]
EC-130E/H	ISR
EC-135N/Y	ISR
F-15A/B	Fighter
F-15C/D	Fighter
F-15E	Fighter
F-16A/B	Fighter
F-16C/D	Fighter
HC-130N/P	Special operations
HH-60G	Helicopter
KC-135E/D/R/T, EC-135C	Tanker
MC-130E	Special operations
MC-130H	Special operations
MC-130P	Special operations
MH-53J/M	Helicopter
RC-135S/U/V/W, TC-135S/W, WC-135W	ISR
T-37B	Trainer
T-38A	Trainer
UH-1N	Helicopter
WC-130H	ISR

[a] ISR = intelligence, surveillance, and reconnaissance.

Table 2.2
Selected F-16C/D Flying DLR Data, FY03Q4

Command	FSC	NIIN	Descriptor	Net Sales ($)	Net Quantity	Standard Price ($)	Exchange Price ($)
Air Education and Training Command	2840	013410171	Hot section module	13,798,571	15	1,293,369	919,905
Air National Guard	2840	013410171	Hot section module	11,958,761	15	1,293,369	919,905
Air Combat Command	2840	014643957	Rotor, turbine	6,083,555	30	233,752	217,270
Air Education and Training Command	2840	014478547	Hot section module	5,593,018	39	452,447	266,334
Air National Guard	2840	013410175	Hot section module	5,317,002	7	1,807,110	886,167
Air National Guard	2840	013410171	Hot section module	4,599,524	5	1,293,369	919,905
Air Education and Training Command	2840	013410175	Hot section module	4,430,835	5	1,807,110	886,167
Air National Guard	2840	014643957	Rotor, turbine	4,144,608	19	233,752	217,270
Air Combat Command	5865	014951126	Control, receiver	3,926,343	27	498,232	145,420
Pacific Air Forces	5865	014951126	Control, receiver	3,635,503	25	498,232	145,420
Pacific Air Forces	2840	014643957	Rotor, turbine	2,824,508	13	233,752	217,270
Air Combat Command	2840	013410175	Hot section module	2,658,501	3	1,807,110	886,167
Air National Guard	2840	013410175	Hot section module	2,658,501	3	1,807,110	886,167
Air National Guard	1270	012330011	Receiver-generator	2,440,639	67	306,719	38,135
Air Combat Command	2840	014922327	Rotor, fan, gas turbine	1,802,630	10	199,458	180,263
Air Force Reserve Command	2840	013410175	Hot section module	1,772,334	2	1,807,110	886,167
Air Combat Command	1270	012330011	Receiver-generator	1,754,209	45	306,719	38,135
Air Force Reserve Command	2840	014643957	Rotor, turbine	1,520,889	7	233,752	217,270

Table 2.2—Continued

Command	FSC	NIIN	Descriptor	Net Sales ($)	Net Quantity	Standard Price ($)	Exchange Price ($)
Air Combat Command	2840	014506905	Rotor, turbine	1,489,711	9	261,027	186,214
Air National Guard	2840	014506905	Rotor, turbine	1,192,096	6	261,027	186,214
Air National Guard	2840	014579971	Rotor, turbine	1,188,905	7	261,027	169,844
Air Combat Command	5895	014977131	Transponder	1,094,076	10	374,848	109,408
Air National Guard	2840	014566799	Rotor, turbine	1,079,669	15	233,752	71,978
Air Combat Command	2840	013396140	Rotor, compressor	1,004,612	8	215,952	143,516
Pacific Air Forces	1270	012330011	Receiver-generator	991,509	27	306,719	38,135

NIIN, noun descriptor, net sales during quarter, net quantity during quarter, standard price, and exchange price.[4]

Net quantities represents the net number of physical units purchased by the commands during the quarter. It equals the sales quantities minus the return quantities. The established price for reparables purchased by the commands is exchange price; the established price for consumables is standard price. The complete database has a code indicating whether a NIIN is a reparable or consumable.[5]

Notice the significant role played by FSC 2840, gas turbine and jet engine, aircraft, the key engine FSC. Net sales equals the dollar value of the transactions during the quarter. Within this FSC, the hot section module constitutes a substantial portion of net sales.

The net quantities and the standard and exchange prices are also used by SAF/FMC to develop aircraft MDS-command–specific MSD price deflators, which we employ in our analysis. Because an important objective of SAF/FMC is to portray the growth in spares "consumption" by the commands, it is appropriate to construct a deflator using MSD-established prices. To develop these deflators, the "market basket" for each MDS-command is defined as all NIINs with identified transactions during a base fiscal year. These NIINs are valued using base-year exchange and standard price, respectively, for reparables and consumables. For a current year, the same base-year NIINs are valued using current-year established prices. The MDS-command price index then equals the current year–to–base year established price valuation of the base-year NIINs.

For example, suppose all the transactions (net quantity) data are available for FY04 and a price file has been published for

---

[4] The NSN, frequently used to identify parts, is a concatenation of the FSC and the NIIN. When the statistical properties of the models are discussed below, we use the more common term, "NSN."

[5] This code is called the expendability/recoverability/reparability/cost (ERRC) Designator. The following codes are relevant: XD1 and XD2, Expendable Investment Item—Depot-Level Repair; XF3, Expendable Expense Item—Field-Level Repair; and XB3—Expendable Expense Item—No Repair. The XD codes ERRCs represent reparables, and the other codes are consumables. These are discussed in AFTOC-Battelle (2004).

FY05.[6] To calculate the MDS-command indexes, one applies the FY05 established prices to the FY04 NIINs containing transaction information, and divides this summed product by the analogous FY04 established price valuation of these same NIINs.

The actual database used in the analysis is an aggregation of this level of data to the MDS-command-FY level. In other words, to obtain an MDS-command-FY data element such as F-16C/D, Air Combat Command (ACC), FY03, one sums across the NIINs and the quarters.[7]

Table 2.3 displays the number of MDS-command cases, by aircraft mission and fiscal year, that results from this aggregation.

About 28 percent of the data points, following aggregation to MDS-command-FY, are fighters. Cargo aircraft and tankers combined constitute about 30 percent of the data. For fighters, cargo aircraft, and tankers combined there may be sufficient data to estimate mission-specific relationships. However, for the other missions, data limitations indicate that combining the data into a pooled longitudinal database is beneficial.[8]

Figure 2.1 compares net sales of flying DLRs to the commands in then-year and constant dollars, where the SAF/FMC deflators are used to compute the constant dollar totals.[9] Clearly, as shown by the

---

[6] The price file, published by AFMC as D043, is discussed in AFTOC-Battelle (2004).

[7] The rationale for aggregating the data is discussed below.

[8] The 446 data points displayed are those used in the preliminary regression. Based on the analysis of the residual scatter plot, 11 of the data points are identified as outliers and excluded from the analysis. In the aircraft system regressions displayed in the main body of this report, 435 data points are employed.

[9] In this chart, we use FY03 as the base year to emphasize the equality of constant and then-year dollars in FY03. Other charts and tables employ FY04 as the base year. One method of comparing MSD inflation, computed by SAF/FMC, is to compare it with the GDP deflator. From 1998 to 2003, the GDP deflator rose from 0.920 to 1.000. In contrast, the MSD deflator increased from 0.638 to 1.000. GDP deflator, therefore, experienced an 8.7-percent price increase over the period, while the MSD deflator increased by 56.8 percent. Years of particularly significant price increases for the MSD deflator were between FY98 and FY99 (an increase of 10.6 percent) and between FY01 and FY02 (an increase of 22.1 percent). Prices are increased because of the requirement to balance the books within MSD. Within a

Table 2.3
Aircraft Mission Structure of Data by Fiscal Year After Aggregation

Aircraft Mission	FY98	FY99	FY00	FY01	FY02	FY03	Total
Bomber	5	5	5	5	4	4	28
Cargo	16	14	17	15	17	15	94
Fighter	20	20	21	21	21	21	124
Helicopter	9	10	10	10	10	9	58
ISR	6	5	5	5	5	5	31
Special operations	8	8	9	8	8	6	47
Tanker	7	7	7	7	6	6	40
Trainer	4	4	4	4	4	4	24
Total	75	73	78	75	75	70	446

change in relative heights of the constant and then-year dollar bars, significant MSD price inflation occurred between FY98 and FY99 and also between FY01 and FY02. The constant dollar bars indicate that real "consumption" of flying DLRs remained fairly constant beginning FY99.

specified period, the Accumulated Operating Revenue must equal zero, and price increases are required to achieve this. Because both the Program Depot Maintenance and Engine Overhaul lines purchase parts from MSD, this price increase affects a substantial component of materiel purchased during overhaul.

Figure 2.1
Flying DLR Net Sales in Then-Year Versus Constant Dollars

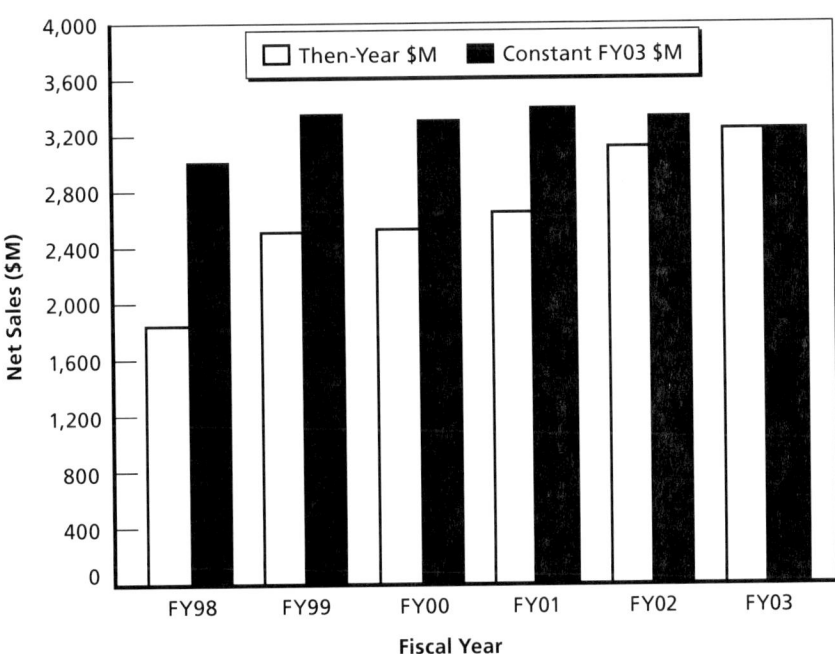

Net Sales Cross-Tabulations

As net sales is the variable that is being explained with the models, several tables summarizing the nature of the data are provided. Aircraft mission type versus fiscal year net sales totals are provided in Table 2.4.

Interestingly, fighters constitute almost 60 percent of net sales over the four-year period. Cargo aircraft constitute about 16 percent of net sales, and bombers constitute about 12 percent of the total. Of course, there is a wide difference in the quantity of aircraft in each category and the total annual flying hours by aircraft type.

Table 2.4
Flying DLR Net Sales by Aircraft Mission Type (FY04 $M)

Aircraft Mission	FY98	FY99	FY00	FY01	FY02	FY03	Total
Bomber	415	380	457	487	539	537	2,814
Cargo	657	754	615	611	651	642	3,930
Fighter	2,032	2,371	2,367	2,471	2,369	2,347	13,957
Helicopter	60	91	98	97	84	85	515
ISR	132	116	104	109	102	97	660
Special operations	123	112	98	127	101	96	657
Tanker	152	165	155	139	146	152	909
Trainer	87	90	94	92	89	83	535
Total	3,658	4,079	3,988	4,133	4,081	4,039	23,978

Table 2.5 shows the command-FY structure of the data. About 30 percent of the relevant net sales over the period are associated with ACC. Air National Guard (ANG) net sales are the second highest among the commands, with about 45 percent of the net sales of ACC, and over three-and-half times those of Air Force Reserve Command. ANG net sales are somewhat larger than Pacific Air Forces and the United States Air Forces in Europe combined. Air Mobility Command (AMC) net sales are about one-third those of ACC.

Table 2.5
Flying DLR Net Sales by Command (FY04 $M)

Command	FY98	FY99	FY00	FY01	FY02	FY03	Total
Air Combat Command	1,328	1,283	1,471	1,543	1,610	1,650	8,885
Air Education and Training Command	453	442	420	441	427	399	2,582
Air Force Materiel Command	N/A	N/A	N/A	0.2	0.3	0.4	1
Air Force Reserve Command	172	199	166	176	189	207	1,109
Air Force Special Operations Command	145	157	150	175	129	149	905
Air Force Space Command	N/A	3	3	3	3	3	15
Air Mobility Command	473	580	459	422	431	402	2,767
Air National Guard	588	711	691	676	632	627	3,925
Pacific Air Forces	314	400	358	439	422	387	2,320
United States Air Forces in Europe	185	304	270	258	238	215	1,470
Total	3,658	4,079	3,988	4,133	4,081	4,039	23,978

Table 2.6 shows the DLR net sales data by aircraft type and subsystem. The FSCs/Groups associated with all the DLR parts have been divided into airframe (and other), avionics, and engine. Airframe (and other) includes aircraft accessories, armaments, and support equipment.

Notice that ISR aircraft are the most avionics intensive of all the aircraft types, with almost 60 percent of their net sales associated with avionics. About 47 percent of bomber and 25 percent of fighter net sales are associated with avionics.

Table 2.6
Flying DLR Net Sales by Subsystem, FY98–FY03 (FY04 $M)

Mission Aircraft	Airframe	Avionics	Engine	Total
Bomber	673	1,399	742	2,814
Cargo	1,461	895	1,574	3,930
Fighter	2,127	3,858	7,972	13,957
Helicopter	299	107	109	515
ISR	157	407	96	660
Special operations	181	262	214	657
Tanker	411	319	179	909
Trainer	138	136	260	535
Total	5,447	7,383	11,147	23,978

For the entire fleet of aircraft, engine net sales are about 43 percent of the total. Fighters are the only aircraft type in which more than 50 percent of net sales are associated with engine.

Other Data Employed

With respect to other data, the basic research philosophy is to use data readily available from official sources in order that the models developed might be used by others participating in Air Force planning activities.

OPTEMPO Data

The operational tempo (OPTEMPO) data employed in this analysis includes flying hours, sorties, and landings. These data are obtained from the Program Data System (PDS). This database, also provided by SAF/FMC, includes information on sorties and landings, and separates the data into flying hours, sorties, and landings conducted during peacetime training and those that occurred during a contingency.

In many analyses, flying hour data are obtained from the Reliability and Maintainability Information System (REMIS). However, the PDS, which starts with REMIS data, contains the final assessment by the commands of Air Force flying hours.

Aircraft MDS Age

Aircraft MDS age data are also obtained from the PDS, and when there are aircraft groupings, a weighted average is computed. It is important to understand that the average age of an aircraft MDS fleet is computed using the age when the tail-numbered specific aircraft is accepted by the Air Force. If a major modification program occurs that changes the MDS designation, the time of acceptance of the tail-numbered aircraft does not change. As a result, the fleet of KC-135Rs has a computed average age based on the year of acceptance of each predecessor KC-135A.

CHAPTER THREE

Specification of Budget Estimating Relationships

We turn now to a discussion of the empirical methodology and model specification. First, let's address the basic specification of the empirical model used in the analysis. Figure 3.1 depicts the structure of the BER that explains these net sales. The explanatory variables can be organized into the following categories: aircraft characteristics, which includes mission type and flyaway cost; OPTEMPO, which represents aircraft usage, measured using flying hours, average sortie duration (ASD), and landings per sortie; and time variables, which include aircraft MDS age and fiscal year categorical variables.

Figure 3.1
Hypothesized BER for Flying DLRs

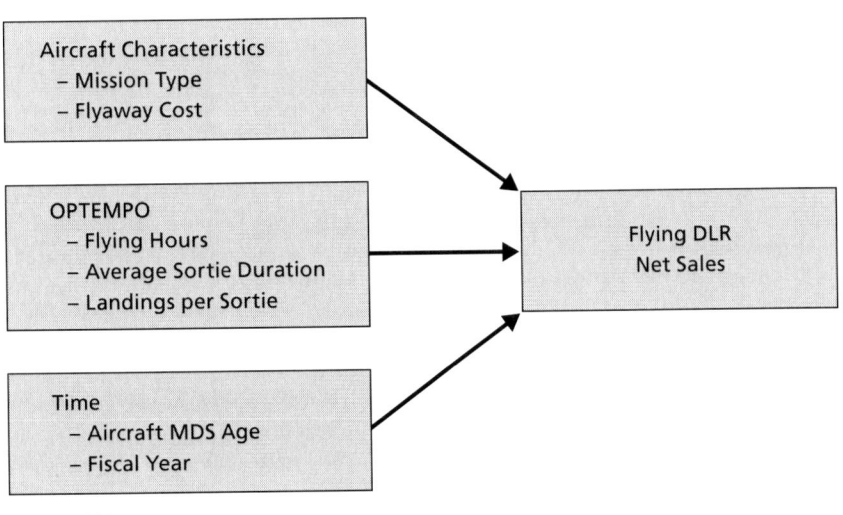

As indicated earlier,

- to estimate the models, we first aggregate the data by individual NIIN-designated parts to obtain MDS-command-FY data points
- there are some combinations of aircraft MDS in the AFCAIG process
- we continue to use the term "aircraft MDS" to describe both an individual aircraft MDS and an SAF/FMC–specified combination of aircraft MDS (e.g., the grouping F-15C/D is referred to as an aircraft MDS)
- a particular data point in the empirical analysis is the annual flying DLR net sales for an aircraft MDS in a particular major command (e.g., F-15C/D flying DLRs in ACC during FY03), and the values of the associated explanatory variables.[1]

Aircraft Characteristics

Using the AFTOC classification system and also several aircraft MDS combinations employed by SAF/FMC, as shown in Table 2.1, the data have been organized into the following broad aircraft mission categories:

- cargo
- bomber
- fighter

---

[1] One may ask why the type of detailed information presented in Table 2.2 is not used directly when building empirical models. One might consider the following factors: (1) One is interested in estimating the effect of different explanatory variables on real consumption, and MDS-command-FY price indexes have been developed at this level; (2) readily available OPTEMPO data are available at the MDS-command-FY-quarter level; (3) age, flyaway cost, and initial operational delivery (IOD) year are available at the aircraft MDS level; and (4) transactions for particular NIINs do not occur in every year, and one would also need a model to predict the probability that a particular NIIN has a transaction, which would then be applied to the observed transactions data.

- helicopter
- ISR
- special operations
- tanker
- trainer.

Flyaway cost is also employed as an aircraft characteristic. As indicated above, these data have been obtained from SAF/FMC.

OPTEMPO

One of the primary purposes of this analysis is to support the AFCAIG process in which CPFH factors are developed. It is natural, therefore, to consider flying hours as the primary indicator of OPTEMPO. However, we also included ASD and landings per sortie as measures of OPTEMPO, that is, of aircraft usage. The combination of flying hours, ASD, and landings per sortie constitute important dimensions of the mission profiles associated with aircraft MDS.

Flying Hours

We have discussed how all the pooled data consist of the various aircraft MDS by command for each fiscal year. A fundamental question concerns the validity of assuming that net sales are proportional to flying hours, which is the standard assumption made during the AFCAIG process. Therefore, different scatter plots are provided for bombers, cargo aircraft and tankers, and fighters. Figures 3.2, 3.3, and 3.4 contain these scatter plots, and, in each graph, a near-proportional relationship between net sales and flying hours is shown.

Figure 3.2
Net Sales Versus Flying Hours for Bombers

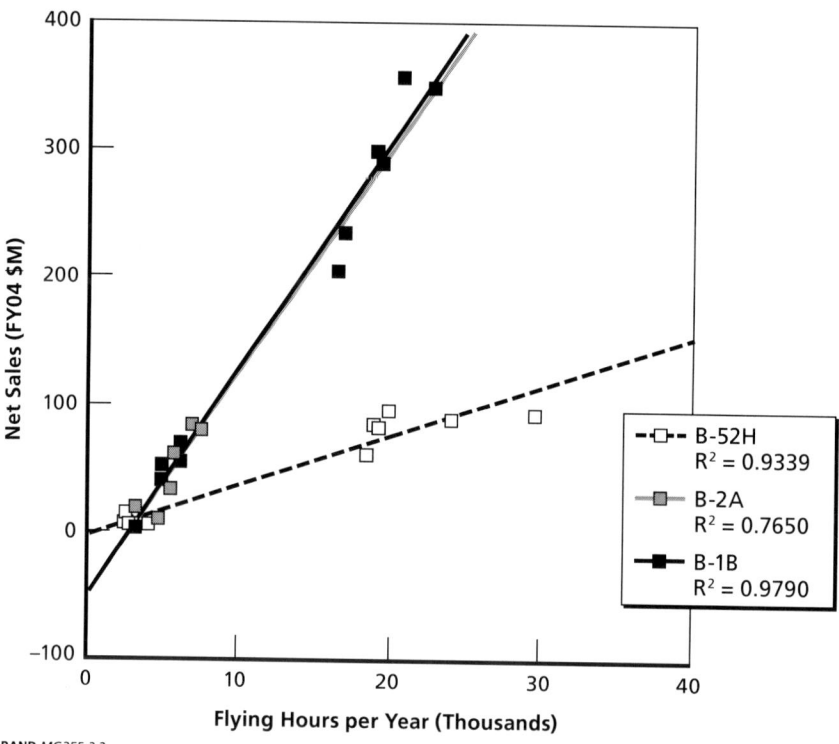

RAND MG355-3.2

Figure 3.2 displays the graph for bombers. Next, in Figure 3.3, we show the graph for cargo/tankers. Finally, Figure 3.4 shows the graph for fighters.

The three figures contain information on the total relationship between net sales and flying hours. While this analysis emphasizes the direct effects of each explanatory variable with the others held constant, at the aircraft MDS level, the proportional effects in the total net sales versus flying hours relationships are pronounced.

Figure 3.3
Net Sales Versus Flying Hours for Cargo/Tankers

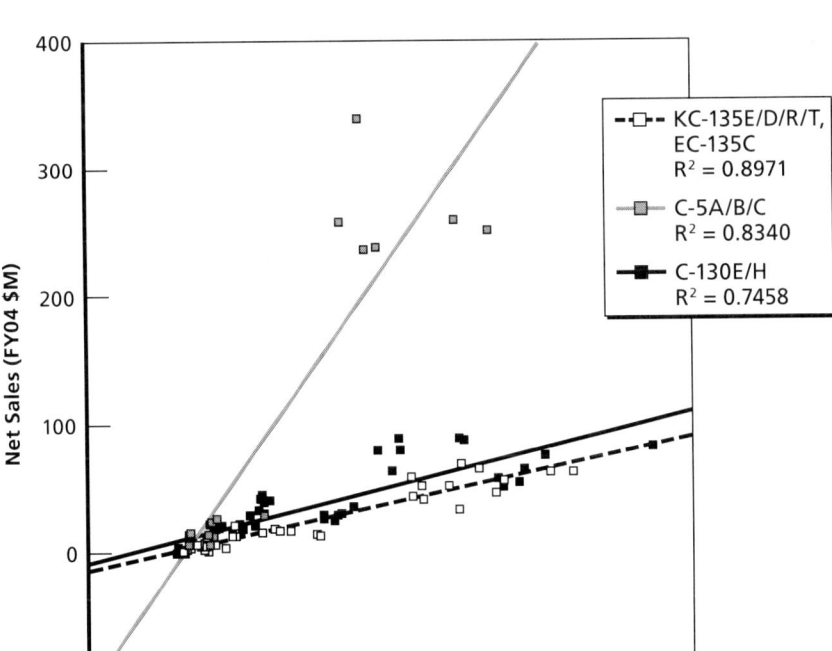

RAND *MG355-3.3*

Another RAND study examined the relationship between repair at the depot and flying hours using monthly data.[2] Using a variety of lag structures, the authors did not find evidence of an association. However, for efficiency reasons, the depot may batch specific NIINs until there are sufficient repair quantities. As a result, the repair may not occur the same fiscal year as the parts are induced into the depot.

---

[2]   See Keating and Camm (2002) for an analysis of the disconnect between flying hours and depot component repair activity and for a discussion of the fixed-cost depot repair elements. Additional discussion of this analysis is in footnote 3 on page 29.

Figure 3.4
Net Sales Versus Flying Hours for Fighters

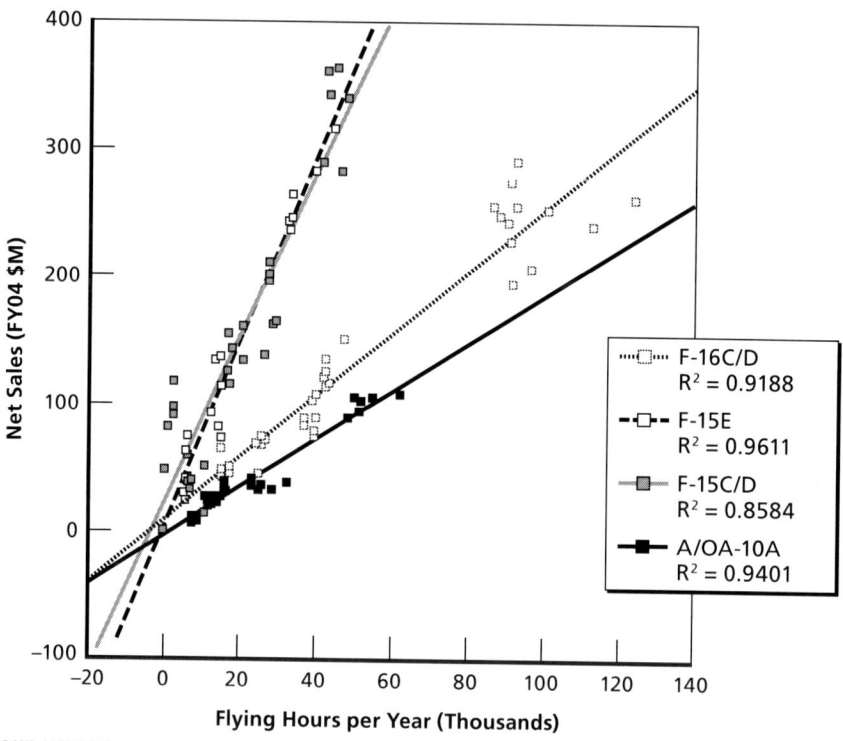

Another issue identified in the aforementioned analysis suggested that there are significant fixed-cost elements in the depot activities that would create a disconnect between total repair costs and flying hours. This may well be the case. The net sales are priced using established prices that include an allocated share of depot overhead. This allocation process may result in accounting costs that differ from the economic costs of repair. These accounting costs appear to support the proportionality hypothesis, and are also the costs used during the budgeting process. Further analysis of the extent to which there are fixed costs, independent of activity levels, is warranted.[3]

There are several other issues that might be raised. For example, it may be argued that budgeting is based on flying hours, and one, therefore, observes net sales following flying hours in accordance with the budgets. However, there are changes between programmed flying hours and the achieved flying hours during the year. We can also view the process as one in which budgets provide the authority to conduct the flying activities, which then generate the net sales to replace the broken parts. Under this view, net sales still depend on flying hours.

Another way of questioning the relationships is to argue that certain costs are allocated using flying hours. This allocation occurs at those bases when several aircraft MDS share the same PEC. However, only a very small portion of the net sales data used for these graphs has been allocated to aircraft MDS using flying hours.[4]

Structures of OPTEMPO Variables

SAF/FMC has developed a classification system, which permits identification of flying hours, sorties, and landings for contingency and peacetime training activities. Contingency flying includes, but is not

---

[3] At the time of the Keating and Camm (2002) analysis, quarterly contracts were employed, which may have attenuated the relationship between flying hours and depot component repair. However, in light of the introduction of the EXPRESS repair priority induction process in DLR maintenance shops, the issue needs to be reexamined. The issue of what depot costs are fixed versus variable with respect to depot repair activities is complex. It is unlikely that there are pure, fixed setup costs independent of activity levels. As the activity level increases, one would expect pressure to be placed on the depot support structure. Eventually it would be optimal to expand this capacity. Conceptually, one is interested in estimating the Long-Run Total Cost function in which all inputs are optimally varied with the activity level.

[4] The allocation occurs when there is a shared PEC at the base level. Because TAI is low for the F-15A/B, F-16A/B, and C-130J, we did not develop scatter plots for these aircraft. For these aircraft, there is limited sharing of PECs, at the base level, with the associated aircraft presented in the charts.

restricted to, the flying done in direct support of Operations Enduring Freedom and Iraqi Freedom.

Table 3.1 displays the flying hours for contingencies and training from FY98 to FY03 for non-CLS aircraft. While about 6.5 to 14 percent of total flying was in support of contingencies from FY98 to FY01, the proportion increased to about 24 and 29 percent, respectively, in FY02 and FY03.

As expected, the contingency flying of trainers did not change much. Although there were increases in contingency flying by fighters and helicopters, bomber, cargo, ISR, special operations, and tanker aircraft experienced a very large increase in such flying activities in FY02

Table 3.1
Contingency Versus Training Flying Hours by Aircraft Mission Type (in thousands)

Aircraft Mission	Type	Flying Hours					
		FY98	FY99	FY00	FY01	FY02	FY03
Bomber	Contingency	0.16	2.55	0.02	0.23	27.08	17.56
	Training	48.23	45.78	49.88	51.26	38.92	37.48
Cargo	Contingency	57.88	70.46	40.54	38.34	132.38	227.95
	Training	330.61	295.02	279.39	284.68	258.77	194.86
Fighter	Contingency	55.40	84.61	31.54	40.34	106.79	115.97
	Training	608.99	568.78	581.02	584.01	562.90	543.52
Helicopter	Contingency	2.88	3.30	1.88	1.79	5.46	7.63
	Training	53.54	53.56	53.41	56.25	51.34	45.38
ISR	Contingency	13.85	17.12	9.59	10.19	29.80	23.15
	Training	33.99	32.47	37.03	39.52	30.33	28.87
Special operations	Contingency	3.61	3.35	1.41	2.03	23.34	17.57
	Training	38.46	38.52	42.27	40.25	29.68	25.77
Tanker	Contingency	25.38	61.74	18.76	25.75	111.90	114.95
	Training	161.21	127.90	139.99	142.38	110.87	112.94
Trainer	Contingency	1.16	1.32	1.51	1.63	1.75	1.61
	Training	322.37	338.99	336.31	340.83	334.47	314.76
Total	Contingency	160.32	244.45	105.25	120.29	438.50	526.40
	Training	1,597.40	1,501.02	1,519.30	1,539.17	1,417.29	1,303.56

and FY03. As expected, the contingency flying did not increase for the trainers. However, it is interesting that training flying decreased for the trainers in FY02 and FY03.

Table 3.2 shows that ASD increased during contingency flying. Overall, peacetime training sorties averaged about 1.7 hours since FY99. However, in FY02 and FY03, ASD during contingency flying increased to 4.3 and 3.7 hours, respectively. Typically, contingency flying hours per sortie in Table 3.2 are always higher for all but the special operations and trainer aircraft, but all fleets except trainers experienced contingency sortie length increases in FY02–FY03.

Table 3.2
Contingency Versus Training Average Sortie Duration by Aircraft Mission Type (hours)

Aircraft Mission	Type	Average Sortie Duration					
		FY98	FY99	FY00	FY01	FY02	FY03
Bomber	Contingency	6.8	12.5	3.3	28.6[a]	12.9	12.0
	Training	5.0	4.8	4.8	4.7	4.5	4.8
Cargo	Contingency	3.0	3.3	3.2	3.3	3.7	3.3
	Training	2.6	2.6	2.6	2.7	2.7	2.5
Fighter	Contingency	2.9	3.5	2.9	2.9	3.6	3.1
	Training	1.5	1.5	1.5	1.5	1.5	1.5
Helicopter	Contingency	1.7	1.6	1.6	1.6	1.8	1.9
	Training	1.6	1.6	1.7	1.7	1.7	1.6
ISR	Contingency	6.9	7.6	7.2	7.3	9.4	9.0
	Training	4.9	5.1	5.0	5.1	4.8	4.7
Special operations	Contingency	2.9	2.4	2.0	2.2	3.5	2.9
	Training	2.8	2.8	3.0	2.8	2.9	2.9
Tanker	Contingency	3.9	5.4	4.9	5.2	5.6	5.5
	Training	3.7	3.8	3.8	3.7	3.8	3.6
Trainer	Contingency	1.1	1.1	1.1	1.1	1.2	1.2
	Training	1.2	1.2	1.2	1.2	1.2	1.2
Weighted Average	Contingency	3.1	3.8	3.3	3.4	4.3	3.7
	Training	1.8	1.7	1.7	1.7	1.7	1.7

NOTE: The weighted average is computed using sorties flown as weights.

[a] The PDS database reports 229 contingency flying hours and 8 contingency sorties for the B-52H in FY01.

Before FY01, the correlation between flying hours and sorties was so high that it was difficult to break out the separate effects in statistical analysis. This close connection was broken in FY02 and FY03, and we examine whether the variation is now sufficiently large as to incorporate additional OPTEMPO variables in an empirical model.

Table 3.3 shows that landings per sortie tend to decline during contingencies. During a contingency, sorties are more likely to be well-defined missions with a single takeoff and landing. Peacetime flying presents a higher proportion of sorties with multiple landings. The change in mission profile in FY02 and FY03 now permits these additional OPTEMPO variables to be included in an explanatory BER.

Table 3.3

Contingency Versus Training Landings per Sortie by Aircraft Mission Type

Aircraft Mission	Type	Landings per Sortie					
		FY98	FY99	FY00	FY01	FY02	FY03
Bomber	Contingency	1.4	1.0	1.8	1.0	1.0	1.1
	Training	2.6	2.7	2.8	2.6	2.8	2.8
Cargo	Contingency	1.0	1.0	1.0	1.0	1.0	1.0
	Training	2.5	2.7	2.8	2.8	2.8	3.0
Fighter	Contingency	1.0	1.0	1.0	1.0	1.0	1.0
	Training	1.0	1.0	1.0	1.0	1.0	1.0
Helicopter	Contingency	2.7	2.2	2.2	2.0	2.3	2.9
	Training	3.1	3.6	3.6	3.5	3.5	3.5
ISR	Contingency	1.1	1.1	1.2	1.1	1.1	1.1
	Training	2.8	3.2	3.9	3.6	3.8	3.6
Special operations	Contingency	1.3	1.4	1.3	1.4	1.1	1.2
	Training	2.5	2.7	2.6	2.6	2.5	2.9
Tanker	Contingency	1.1	1.1	1.1	1.1	1.1	1.1
	Training	3.0	3.4	3.4	3.2	3.3	3.4
Trainer	Contingency	1.7	1.9	2.0	1.7	1.5	1.5
	Training	2.3	3.0	3.0	2.9	2.9	2.8
Weighted Average	Contingency	1.1	1.1	1.1	1.1	1.1	1.1
	Training	1.8	2.1	2.1	2.1	2.1	2.0

NOTE: The weighted average is computed using sorties flown as weights.

A complication in developing an empirical model that captures the effect of contingencies is that there does not exist a well-defined data series for flying DLR net sales during a contingency that also accounts for the relationship between the demand for parts during a contingency and the nature of the flying activities before the contingency. Also, contingency flying can be expected to have an effect on demands for some period subsequent to the contingency. As a result, the annual DLR net sales data employed in the analysis combine both contingency and training costs. This should permit one to capture the full effect of a change in the OPTEMPO variables on net sales during a fiscal year.[5]

Time Variables

There has been significant discussion of the effect of aging on net sales. However, aging occurs in time, and other factors affecting net sales are related to time. Therefore, it is necessary to examine the different "fiscal year effects" in order to determine whether they influence the aging effect. In addition, fiscal year categorical variables may capture variables excluded from the model that are correlated with these variables. If the fiscal year variables are not significant, this increases one's confidence that the certain important missing variables have not been excluded from the analysis.

Associations Between Net Sales and Aircraft MDS Age

First, we examine the direct associations between net sales and aircraft MDS age. One of the difficulties associated with estimating the aging

---

[5] Using SRAN information as well as mission codes, one can identify transactions at deployment bases during contingencies. An additional issue concerns the Readiness Support Packages used during contingencies. Although the budgeting issues are complex, a survey of the Air Force commands indicates that, during contingencies, the flying units continue to pay the established prices for parts demanded. However, there may be subsequent reimbursement for costs incurred, particularly for ANG.

Figure 3.5
Air Force–Wide Net Sales for Bombers Versus MDS Age, FY98–FY03

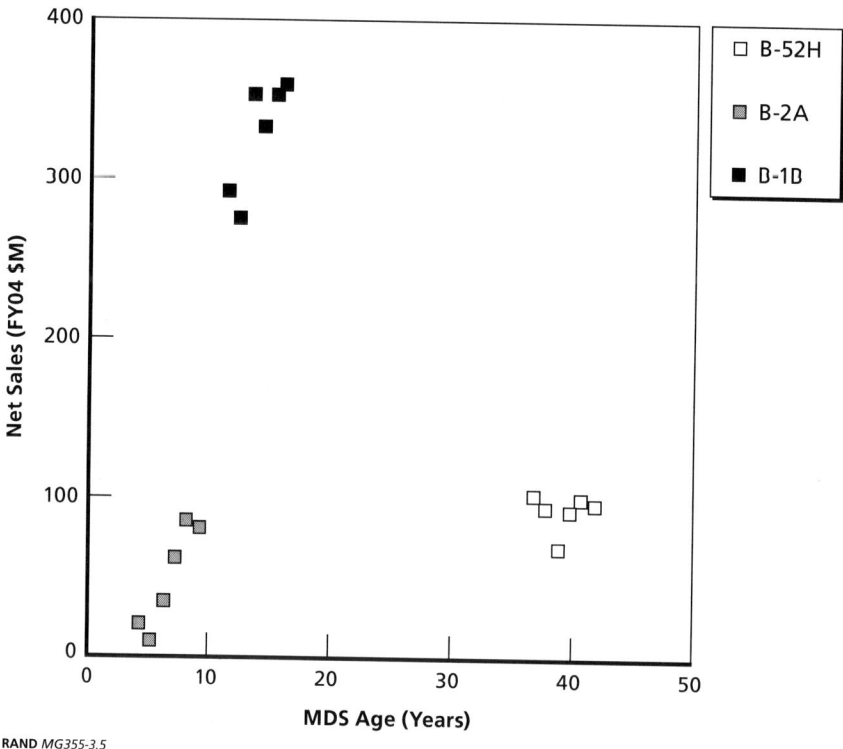

effect is that other variables that are correlated with aircraft age can affect net sales. Therefore, examination of the total relationship between net sales and aircraft MDS age does not demonstrate whether an aging effect is present. However, just as we illustrated some of the associations between net sales and flying hours, it is appropriate to illustrate the association between net sales and aircraft MDS age using scatter plots.

Because published aircraft age computations are at aircraft MDS level, and command-level aircraft age is not readily available, we present Air Force–wide scatter plots for the period FY98–FY03. The graph for bombers is displayed in Figure 3.5.

We obtain a positive association between net sales and age for the B-1B and the B-2A. There is no apparent association for the B-52H.

Figure 3.6
Air Force–Wide Net Sales Versus Age for Cargo/Tankers, FY98–FY03

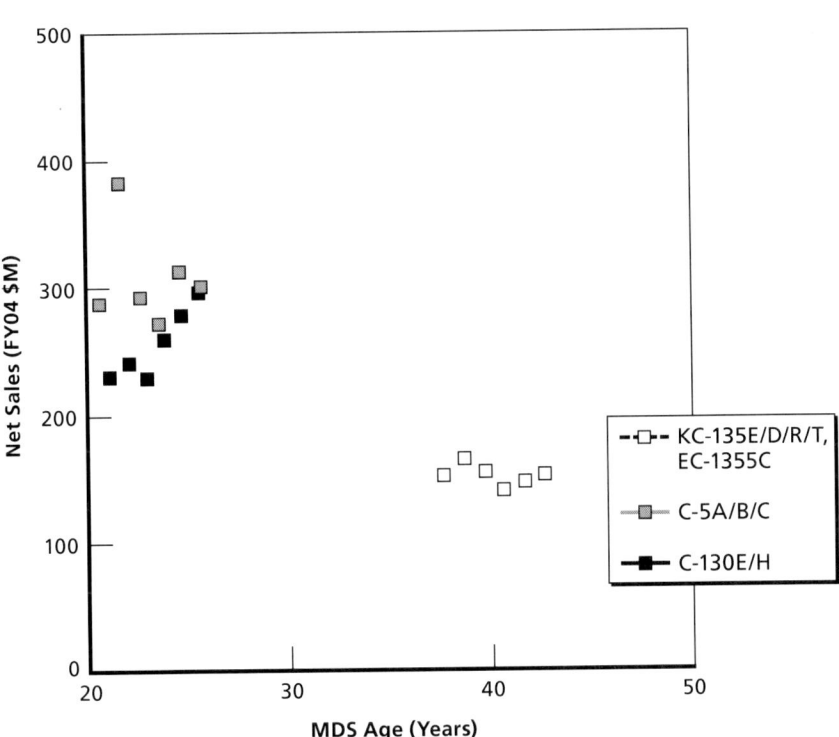

Next, selected cargo/tankers are depicted in Figure 3.6. Only the C-130E/H shows a clear positive association between net sales and age.

Figure 3.7 contains the graph for fighters. Positive associations are apparent for the F-15C/D, F-15E, and F-16C/D.

It is important to reiterate that these types of plots are inappropriate for identifying the direct effect of aging on net sales, other things equal. In the background of these plots, other factors correlated with age affect net sales. To estimate the aging effect, one must account for the fact that such variables as flyaway cost and flying hours, which are correlated with MDS age, also affect the cost of DLRs. It is necessary, therefore, to control for these variables when estimating the aging effect.

Figure 3.7
Air Force–Wide Net Sales Versus Age for Fighters, FY98–FY03

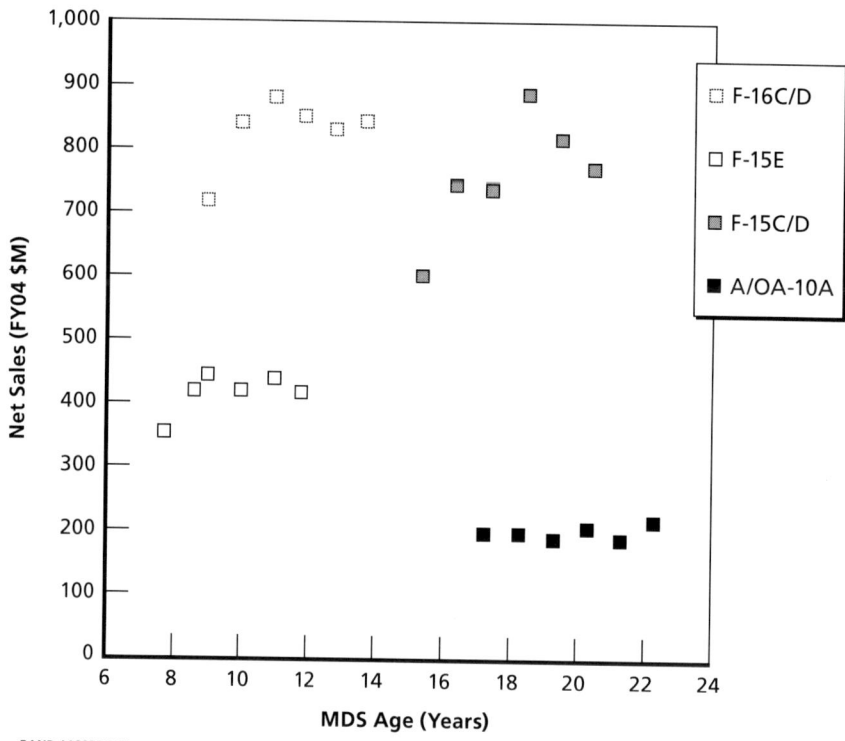

It may be helpful if we examine various factors that impede one's ability to estimate the direct aging effect. For example, while we can control for readily measurable variables that interact with aircraft age, the interplay between aircraft modifications and aging presents a particular challenge. At present, there does not exist the necessary modification series by aircraft MDS that begins at the IOD year. This hinders one's ability to determine how replacement cost has changed from flyaway cost as a result of the modifications. Figure 3.8 illustrates some of the broad features of this complicated area.[6]

---

[6] This figure is a variation of Figure 3.1, first presented in Pyles (2003).

Figure 3.8
Theoretical Direct and Indirect Effects of Aircraft Aging on Costs

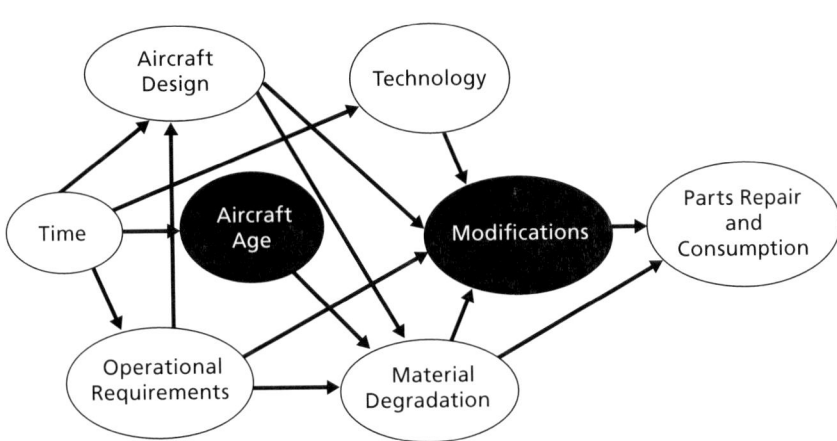

RAND *MG355-3.8*

Material degradation of a particular aircraft design can be caused by both aircraft aging and changes in the operational environment. Over time, as the equipment ages, aircraft may operate in more-demanding environments as the threat changes. As missions evolve and sortie rates change, harsher climates are encountered that can, in turn, affect the demand for parts.

Both aging and operational environment changes, therefore, can directly increase parts consumption and repair, and also generate a demand for new modifications. The cost of these modifications is affected by both the extent of the material degradation and by the level of technology that has been achieved. Changes in the operational environment also result in new modifications that are designed to increase the capability of the aircraft.

Modifications also affect parts repair and consumption, particularly when the capability of the aircraft is enhanced. As a result, the total effect of both aging and changes in operational environment on parts repair and consumption is directly affected by material degradation, indirectly affected by this degradation through its effect on aircraft modifications, and indirectly affected by the modifications that are directly caused by changes in the operational requirements.

There may, therefore, be a separate operational environment effect, over and above the aging effect, that affects parts repair and consumption. The changes in parts repair and consumption may occur either directly, as a result of material degradation, or, indirectly, as a result of modifications.

The reason it is difficult to isolate the separate effects of aging and changes in operational requirements on net sales is that both effects take place in time. Some part of the change in operational requirements may, therefore, be perfectly correlated with the age of the aircraft. If this occurs, the effect of such a change on net sales would be captured as part of the estimated aging effect.

However, there may be aspects of the changing operational environments and associated requirements that are not perfectly correlated with age. These changes are likely to be very difficult to measure empirically, although there may be resulting changes in the OPTEMPO and other measured variables. In addition, categorical variables for each relevant data year can capture the effects of changes in the operational environment and aid the identification of a separate aging effect. Therefore, the determination below—that these fiscal year variables are not significant—is an important finding.

It is, however, possible that measurable variables such as the OPTEMPO variables, or the fiscal year of the flying activity, do not fully capture the effects of changes in the operational environment. If these changes in the operational environment are correlated with aging, part of their effect on net sales would be captured by the measured effect of aircraft aging.

Figure 3.8, therefore, suggests that precisely measuring the pure effect of aging on net sales is quite challenging. As seen below, we are able to isolate an aging effect using a statistical model that carefully controls for many of the key variables that also affect the demand for parts. The standard qualification that applies for variables correlated with aircraft age that are missing from the statistical model applies. Yet, even if we are not fully successful in isolating the pure aging effect, the explanatory power of the models described below permits our results to be used as an aid in Air Force acquisition planning and budgeting.

Figure 3.8 should also convey the fact that part-level summaries of maintenance man-hours per flying hour, over time, may not fully capture the complexities associated with estimating the effect of aircraft age on net sales. The direct aging effect that we estimate is best interpreted as the effect of material degradation, technology enhancements, and modifications on parts repair and consumption that is not captured by those variables included in the analysis that are directly affected by changes in operational requirements.

CHAPTER FOUR

Empirical Findings

Functional Form of Specified Model

As illustrated graphically in Figure 3.1, we initially employ the following functional form in our hypothesized initial model:[1]

$$Ln(\text{Net Sales}) = \beta_0 + \beta_1 \text{ Aircraft Mission Effects} + \beta_2 \text{ Ln(Flyaway Cost)}$$
$$+ \beta_3 \text{ Ln(Flying Hours)} + \beta_4 \text{ Ln(Average Sortie Duration)}$$
$$+ \beta_5 \text{ Ln(Landings per Sortie)} + \beta_6 \text{ MDS Age}$$
$$+ \beta_7 \text{ Fiscal Year Effects} + U. \tag{4.1}$$

The aircraft mission effects variable refers to the categorical variables bomber, cargo, helicopter, ISR, special operations, tanker, and trainer, each with its own coefficient, where fighters is the reference variable. For simplicity, we have indicated the "coefficient" of these effects as β_1. The time variables in the base model are aircraft MDS age and the fiscal year variables, which for simplicity are designated β_7.

Several comments about the functional form used in Equation (4.1), which might be described as a log-log-linear model, are appropriate. One would likely expect the "true" response schedule representing the effect of the explanatory variables on net sales to be nonlinear with

---

[1] For simplicity, we suppress the subscripts of Equation (4.1). The variables—net sales, flying hours, and ASD—are measured using the ith MDS, jth command, and tth time period. The aircraft mission effects and flyaway cost are defined for the ith MDS, and are not dependent on command or fiscal year. The variable MDS age depends on the ith MDS and tth fiscal year, but is not dependent on command. The fiscal year variables are associated with the tth fiscal year.

significant interaction among the variables. By interaction, we mean, for example, that the effect of flyaway cost on net sales depends on the age of the aircraft. For the variables net sales, flyaway cost, and flying hours, the functional form used in Equation (4.1) is a first-order approximation (in the logs); a linear approximation is used for the remaining variables.

The estimated coefficients of this type of model are easy to interpret. For example, β_2 and β_3 represent the percentage change in net sales resulting from a 1-percent change in flyaway cost and flying hours, respectively. A similar interpretation is given to the coefficients of ASD and landings per sortie. The coefficient β_6 equals the proportionate change in net sales when there is a one-year change in MDS age. The categorical variables—aircraft mission type and fiscal year effects—are included in the traditional linear form.

Empirical Estimation

The traditional method of estimating the specified model is ordinary least squares (OLS). This method selects, as the estimated coefficients, those that minimize the sum of the squared differences between the actual and predicted values. The actual minus the predicted values are called the residuals of the model. As discussed further in the appendix, the residuals, which reflect the model's unobservable errors, originally consist of 446 aircraft MDS-command-FY data points. However, initial analysis indicates that 11 data points, from selected aircraft MDS, should be removed from the analysis. These data points have highly negative residuals in the exploratory statistical model. Table 4.1 contains the results obtained when we estimate the model represented in Equation (4.1).

We find that none of the fiscal year variables are statistically significant. This provides support for removing these variables from the analysis. This nonsignificance also provides evidence that missing variables correlated with these fiscal year variables are not excluded from the analysis.[2]

The revised model is presented in Table 4.2. This model provides the primary results of this analysis.

We find that, other things equal, the bomber, cargo, helicopter, ISR, special operations, tanker, and trainer all have significantly lower net sales than fighters.

Table 4.1
Budget Estimating Relationship—Specified Model with Fiscal Year Variables, FY98–FY03

Explanatory Variable	Coefficient	t-Statistic	Significance
(Constant)	5.045	12.569	0.000
Bomber	−1.252	−4.991	0.000
Cargo	−1.996	−11.373	0.000
Helicopter	−0.823	−4.762	0.000
ISR	−1.917	−7.484	0.000
Special operations	−1.282	−7.262	0.000
Tanker	−2.692	−11.287	0.000
Trainer	−1.630	−8.264	0.000
Ln(Flyaway Cost)	0.819	14.108	0.000
Ln(Flying Hours)	1.038	33.743	0.000
Ln(Average Sortie Duration)	−0.327	−1.996	0.047
Ln(Landings per Sortie)	0.199	1.899	0.058
MDS age	0.029	5.851	0.000
FY99	0.040	0.377	0.706
FY00	0.036	0.347	0.729
FY01	−0.002	−0.016	0.987
FY02	−0.151	−1.411	0.159
FY03	−0.115	−1.051	0.294

NOTE: Dependent variable: Ln(Net Sales), FY04 MSD$.
$R^2 = 0.851$; $n = 435$; standard error of the estimate (SEE) = 0.636.

---

[2] An F-test was also used to test the null hypothesis that at least one of the fiscal year variables are statistically significant against the alternative hypothesis that none of them is significant. The F-test statistic equals 1.119, which can be compared with an F-statistic associated with a 0.01 level of statistical significance approximately equal to 3.04. Therefore, we can confidently accept the null hypothesis.

Table 4.2
Budget Estimating Relationship Model 1: Without Fiscal Year Variables, FY98–FY03

Explanatory Variable	Coefficient	t-Statistic	Significance
(Constant)	5.104	12.797	0.000
Bomber	−1.197	−4.811	0.000
Cargo	−1.970	−11.332	0.000
Helicopter	−0.855	−4.970	0.000
ISR	−1.863	−7.341	0.000
Special operations	−1.254	−7.147	0.000
Tanker	−2.631	−11.179	0.000
Trainer	−1.643	−8.351	0.000
Ln(Flyaway Cost)	0.814	14.065	0.000
Ln(Flying Hours)	1.035	33.723	0.000
Ln(Average Sortie Duration)	−0.371	−2.284	0.023
Ln(Landings per Sortie)	0.229	2.216	0.027
MDS age	0.027	5.615	0.000

NOTE: Dependent variable: Ln(Net Sales), FY04 MSD$.
$R^2 = 0.849$; $n = 435$; SEE = 0.637.

The results also show that a 1-percent increase in flyaway cost increases net sales by about 0.81 percent. This means that flyaway costs result in less than proportional increases in net sales. Aircraft consist of DLRs in addition to airframe structure and complete engines. The flyaway cost effect identifies the part of flyaway cost that is associated with net sales, other things equal.

With respect to the OPTEMPO variables, we find that a 1-percent increase in flying hours increases net sales by about 1.04 percent. A working assumption of the AFCAIG process is that costs are proportional to flying hours. This result indicates that a near-proportional relationship holds when other variables are held constant. In Figures 3.2, 3.3, and 3.4, we saw that a near-proportional direct relationship holds as well for bombers, cargo/tankers, and fighter aircraft.

A 1-percent increase in ASD reduces net sales by about 0.37 percent. DLR cost declines when sortie duration increases. Although it may appear that this result might be confounded with contingency flying when ASD is higher, RAND work in progress obtains a similar result for commercial aircraft costs.[3]

We also find that a 1-percent increase in landings per sortie increases net sales by about 0.23 percent. Increases in landings per sortie stress the aircraft and its DLRs, so the result is not surprising. As shown in Table 3.3, landings per sortie decline during contingencies, so there is an additional cost-reducing effect.

A one-year increase in aircraft MDS age increases net sales by about 2.7 percent. We believe this results from both material degradation and the increase in cost associated with capability modifications being added to the aircraft. There could also be a technology component associated with this result. To the extent that aircraft with an earlier IOD year have higher net sales than aircraft of more-recent vintage, the aging effect may embody such technical progress.

Figure 4.1 contains a scatter plot of the dependent variable Ln(Net Sales) versus the predicted value of the dependent variable. The diagram also displays the $R^2 = 0.849$ contained in Table 4.2.

---

[3] We examined a model in which flying DLRs were related to aircraft mission, flying hours, and the interaction between flying hours and command. The interaction variables were not statistically significant, and flying hours were nearly proportional to net sales. This model supports the near-proportionality conclusion identified in Table 4.2 and in the graphs displayed in Figures 3.2, 3.3, and 3.4. The work, in progress, that identified a negative ASD effect for commercial aircraft is being conducted for engine-related O&S costs by Mike Boito and Greg Hildebrandt.

Figure 4.1
Scatter Plot of Ln(Net Sales) Versus Predicted Ln(Net Sales)

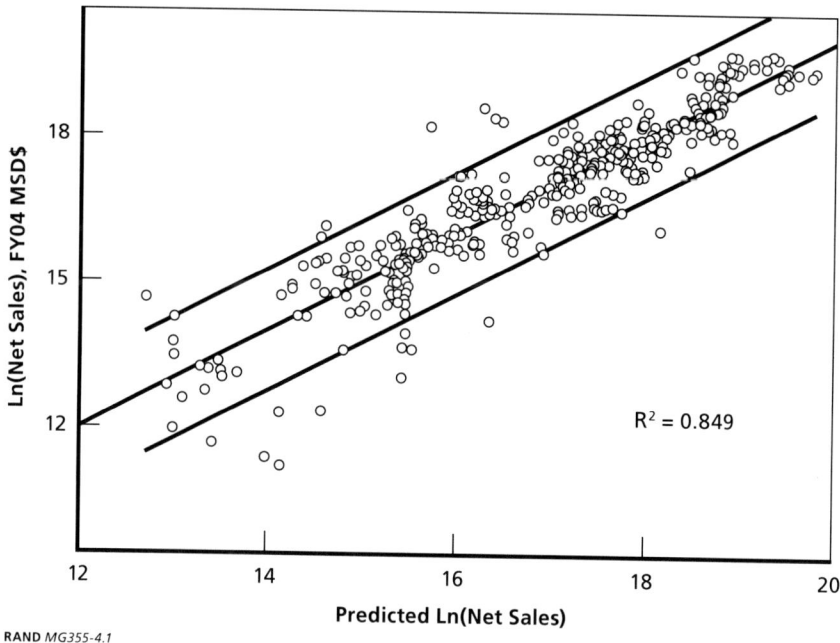

RAND *MG355-4.1*

The indicated bands constitute the 95-percent confidence interval of the individual prediction, which in this case is the Ln(Net Sales)
for a specified MDS-command-FY combination with predicted values
within the range displayed on the chart.

The chart suggests that the most observations above a predicted
value of about 15 fall within the confidence bands for the individual prediction. However, for lower predicted values, there is a greater
tendency for the observations to fall outside the confidence bands. At

higher predicted values, the values of Ln(Net Sales) tend to be clustered inside the confidence bands.[4]

Prediction Model Including Serial Correlation

As discussed, the policy response models focus on contemporaneous causality and do not account for possible interactions of variables over time. Certainly, one must consider the possibility that causal effects occur across time, even if it is not possible to fully explain the nature of the intertemporal effects. Therefore, we examine this phenomenon by considering a model in which there is serial correlation among the residuals across time. For simplicity, we constrain the serial correlation coefficient to be the same across each MDS-command combination. This is one of the simplest, but also most widely used, intertemporal structures.

The error term of Equation (4.1) is assumed to possess the following structure:[5]

$$U = \rho U_{-1} + V. \qquad (4.2)$$

Table 4.3 displays the results obtained.

---

[4]   This tendency for observations with lower predicted Ln(Net Sales) may suggest evidence of heteroscedasticity; and several weighting variables, including the number of NIINs associated with an MDS and predicted net sales, were examined. The indication is that few data points with large values for the lower values of predicted Ln(Net Sales) dominate the effect of the different weighting variables. We decided to retain the use of OLS because of the method's robust statistical properties. As we have already eliminated 11 outliers from the analysis, it is likely inadvisable to eliminate a second set of outliers based on the revised regression. An additional consideration is that, when the serially correlated residual model is used, weighting the data using Predicted Ln(Net Sales) does not have a significant effect on the estimated coefficients.

[5]   The subscripts of the first-order autoregressive process are i and j, representing MDS and command. Therefore, each MDS-command combination has the same estimated value of ρ. It is approximately equal to the serial correlation between the value of U and fiscal year = t and fiscal year = t - 1.

Table 4.3
Budget Estimating Relationship Model 2: With Serial Correlation, FY98–FY03

Explanatory Variable	Coefficient	t-Statistic	Significance
(Constant)	6.282	8.700	0.000
Bomber	−0.948	−2.170	0.032
Cargo	−1.919	−6.544	0.000
Helicopter	−1.286	−4.432	0.000
ISR	−1.779	−4.437	0.000
Special operations	−1.365	−4.411	0.000
Tanker	−2.507	−6.420	0.000
Trainer	−1.711	−4.413	0.000
Ln(Flyaway Cost)	0.781	6.926	0.000
Ln(Flying Hours)	0.944	17.349	0.000
Ln(Average Sortie Duration)	−0.632	−4.139	0.000
Ln(Landings per Sortie)	0.402	3.233	0.001
MDS age	0.022	2.596	0.011

NOTE: Dependent variable: Ln(Net Sales), FY04 MSD\$.
Calculated R^2 = 0.947, n = 356; ρ = 0.845; Standard Error (ρ) = 0.023.

Note first that the serial correlation coefficient, ρ = 0.863, is fairly large and statistically significant. Significant residual correlation across time is now being captured.

If we compare this model with that reported in Table 4.2, we see that all aircraft mission variables are negatively statistically significant, as before. Given the size of ρ, the coefficients of Ln(Flyaway Cost) and Ln(Flying Hours) are similar. However, Ln(Average Sortie Duration) is more negative and Ln(Landings per Sortie) is more positive than the results presented in Table 4.2. It is possible that this occurs because the significant change in these two variables occurred in FY02, and this change represented a shock to the Flying Hour Program that took some time to work itself out. It is frequently argued that shocks of this type are the source of serial correlation. Therefore, there may be interaction between ρ and the coefficients of these two OPTEMPO variables.

Interestingly, while the aging effect declined from 2.7 percent per year of age in Table 4.2 to 2.2 percent in Table 4.3, the variable remains statistically significant. The fact that the fiscal year variables and serial correlation do not interact significantly with the time variable adds support to the fact that aging is a real phenomenon. All three variables are time related, but neither fiscal year nor serial correlation eliminate the aging effect.

It is also appropriate to depict the scatter plot between Ln(Net Sales) and the Predicted Ln(Net Sales), where the latter incorporates the effect of serial correlation. This graph is presented in Figure 4.2.

Figure 4.2
Scatter Plot of Ln(Net Sales) Versus Predicted Ln(Net Sales) with Serial Correlation Adjustment

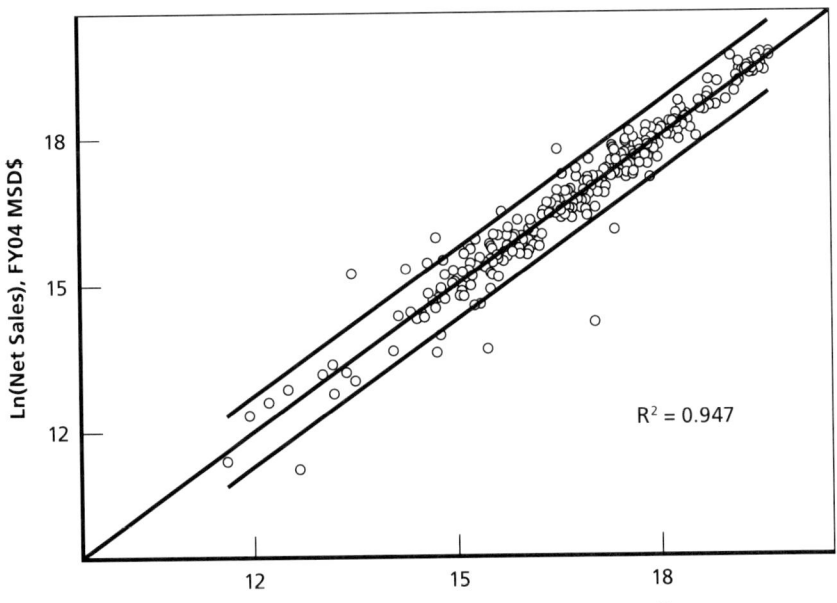

The correlation between Ln(Net Sales) and the predicted value has increased from 0.849 to 0.947 as a result of accounting for serial correlation. Also notice that fewer data points lie outside the outer 95-percent confidence interval.[6]

This model, clearly, can play a fruitful role in forecasting net sales. However, as mentioned, there are many possible forms of correlation structures that might be investigated and we have used the simplest. The use of this serial correlation adjustment represents a method of accounting for dependency across time without attempting to fully model the intertemporal structural relationships.

Table 4.2, which portrays contemporaneous causality, represents our best estimate of the effect of varying one of the structural variables, that is, in conducting policy response analysis. However, Table 4.3 portrays a model that has substantial predictive power. We believe both models are relevant: One provides structural explanation that is particularly useful in policy response analysis, and the second permits effective forecasting.

---

[6]   As indicated in footnote 4 on page 47, this supports retaining homoscedasticity as an assumption rather than making an adjustment to correct for possible heteroscedasticity. Also, scatter plots were developed for all the aircraft mission categories. With the exception of special operations aircraft, all of the R^2 were quite high. The following R^2 were obtained: bombers, 0.926; cargo: 0.976; fighters: 0.924; helicopters: 0.870; special operations, 0.661; tankers, 0.867; and trainers, 0.959.

CHAPTER FIVE

Conclusions

The primary analytical results of this study are reported in Chapter Four, where, after eliminating the nonsignificant fiscal year variables, we present two BERs following the elimination of the fiscal year variables. The first, called Model 1, displayed in Table 4.2, employs only contemporaneous data. The second, Model 2, represented in Table 4.3, takes into account the serial correlation found among the intertemporal residuals.

Figure 5.1 summarizes the findings for these two BERs.

The two BERs contain reasonably similar results for the common variables. We find that non-fighter aircraft missions have lower net sales. The effect of a 1-percent change in flyaway cost, or flying hours, or a one-year change in aircraft MDS age or IOD year have similar percentage effects in the two models. Model 2 does show somewhat higher effects for ASD and landings per sortie. This may be associated with the interaction between these two variables and the serial correlation coefficient, ρ.

As we have seen, during contingencies, ASD increases and landings per sortie decline. Both of these changes in mission profiles act to reduce net sales. Why this occurs during contingencies requires further analysis. Increases in ASD do decrease commercial maintenance cost, suggesting that this is a phenomenon that extends beyond contingency flying. One may hypothesize, however, that that there may be deferred maintenance activities during contingencies. Another hypothesis is that more maintenance takes place in the backshops during contingencies.

Figure 5.1
Comparison of Two Budget Estimating Relationships with Different Error Structures

Model 1: Effect of Contemporaneous Variables on Flying DLR Net Sales	**Model 2: Effect of Contemporaneous Variables and Serially Correlated Error Term on Flying DLR Net Sales**
• Non-fighter aircraft missions have significantly lower net sales	• Non-fighter aircraft missions have significantly lower net sales
• 1-percent change in variable has indicated significant percentage effect on net sales	• A 1-percent change in variable has indicated significant percentage effect on net sales
– Flyaway cost: 0.81	– Flyaway cost: 0.78
– Flying hours: 1.04	– Flying hours: 0.94
– Average sortie duration: –0.37	– Average sortie duration: –0.63
– Landings per sortie: 0.23	– Landings per sortie: 0.40
• A one-year change in variable has indicated percentage effect on net sales	• A one-year change in variable has indicated percentage effect on net sales
– Aircraft MDS age: 2.7	– Aircraft MDS age: 2.2
• $R^2 = 0.85$	• $\rho = 0.85$
	• $R^2 = 0.95$

It must be recognized, however, that we have not used net sales data that have been identified with contingency flying. This requires extensive analysis of the net sales data and would not, in any event, by itself, take account of deferred maintenance. Nevertheless, because annual data are employed, the methodology does capture some of the impact of contingency flying on net sales both before and following deployment. However, a more-complete analysis of contingency costs is required before any firm observations can be made.

Our approach has been to focus on explanatory causal models that rely on contemporaneous data. The structural models we present are consistent with the causal hypotheses. While it is typically not possible to prove causality using observational data, which is an issue that pervades social science research, the results obtained provide evidence that the structural models estimated can be used for the intended purposes.

Finally, it is appropriate to mention that initiatives are under way that will attempt to change Air Force maintenance concepts and attenuate some of the MSD price increases associated with the AFWCF. For example, the Centralized Asset Management initiative is examining, among other issues, the return to the free issue of spare parts. Any major change in the accounting business rules used by the Air Force will require this analysis to be revisited.

APPENDIX

Additional Empirical Results

Empirical Strategy

This appendix expands the discussion of the empirical analysis. In the main body of the text, we present results obtained using OLS multiple-regression technique, including the use of the model that contained a serially correlated error process. In this appendix, we discuss the issue of outliers, and effect of outlier removal on the significance of fiscal year variables. To further verify the existence of a well-defined aging effect, we also estimate a model in which first differences are computed. This model eliminates variables that are constant across individual MDS-command combinations that occur each fiscal year, including aircraft mission and flyaway cost. Within an aircraft MDS, the effect of command over time is also removed.

Fiscal Year Effects

Fiscal year variables must be considered in an analysis of this sort to increase confidence in the empirical results. Because the average age of the Air Force fleet has been increasing over time, one needs to be certain that other factors, which affect cost over time, are not given an aging-effect imputation. For example, MSD prices may rise for reasons other than MDS age, perhaps to cover a budget deficit in the AFWCF. The use of the appropriate price deflator can address this, but perhaps only partially. Also, a fiscal year variable can capture any changes in the scope of the data used in the analysis. Other factors that may be captured by a fiscal year variable include

- changes in operational requirements including contingencies, deployments, and the command structure of activities
- modifications unassociated with age
- changes in maintenance concepts
- changes in reimbursement practices (e.g., Propulsion Business Adjustment).

It is simply not possible to incorporate in a statistical model all the factors that can affect net sales over time. Either data are not available, or a factor is very difficult to measure. Yet numerous factors, other than those accounted for in the model, can affect net sales. The fiscal year variables determine whether such factors have an important impact on the results and provide additional confidence that an aging effect is, in fact, being estimated. Also, if we find that the fiscal year variables are not statistically significant, there is greater confidence that important variables that affect cost, and which are associated with particular fiscal years, are not inadvertently excluded from the analysis.

We start with an MDS-command-FY–based dataset that contains 446 observations.

Table A.1 contains the preliminary OLS regression results when all 446 data points are employed.

Figure A.1 shows the residual plot of the standardized residuals from this model versus the standardized predicted value for the aircraft MDS-command-FY data points.

Table A.1
Preliminary Ordinary Least Squares Regression

Explanatory Variable	Coefficient	t-Statistic	Significance
(Constant)	4.248	7.609	0.000
Bomber	−1.115	−3.132	0.002
Cargo	−2.015	−8.115	0.000
Helicopter	−1.003	−4.139	0.000
ISR	−1.812	−4.983	0.000
Special operations	−1.380	−5.535	0.000
Tanker	−2.777	−8.209	0.000
Trainer	−1.785	−6.338	0.000
Ln(Flyaway Cost)	0.821	10.123	0.000
Ln(Flying Hours)	1.112	25.663	0.000
Ln(Average Sortie Duration)	−0.348	−1.519	0.129
Ln(Landings per Sortie)	0.025	0.172	0.864
MDS age	0.041	6.284	0.000
FY99	0.040	0.267	0.790
FY00	−0.064	−0.431	0.667
FY01	−0.131	−0.872	0.384
FY02	−0.361	−2.392	0.017
FY03	−0.387	−2.511	0.012

NOTE: Dependent variable: Ln(Net Sales), FY04 MSD$.
$R^2 = 0.770$; $n = 446$; SEE = 0.911.

There are 11 data points removed from the data set that cannot be adequately explained by the model. These include data with negative standardized residual values (less than −3.0) and C-130J aircraft possessed by ANG. From 1999 through 2003, two of these ANG C-130J data points could not be incorporated in the analysis, and the standardized residuals for FY00, FY02, and FY03 were −2.70, −3.20, and −2.55, respectively. This reduces the usable data from 446 to 435 data points.[1]

---

[1] In addition to the ANG C-130Js, the excluded cases are the Air Education Training Command HH-60G in FY00, FY01, FY02, and FY03; the ANG MC-130P in FY01, FY02, and FY03; and the AMC UH-1N in FY03.

Figure A.1
Residual Plot from Preliminary Regression

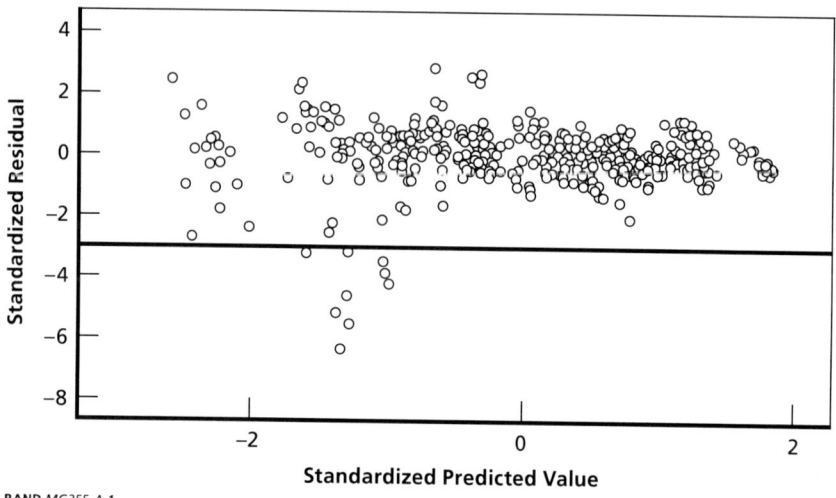

Table 4.1 (see p. 43) shows the results obtained when OLS is used to analyze the 435 remaining data points, and all the fiscal year variables are included. None of the fiscal year variables are statistically significant, and these are removed from the analysis. Table 4.2 (see p. 44) displays the empirical results when these variables are removed.

First-Difference Model

As indicated, each observation is a particular aircraft MDS-command-FY combination. This data combination constitutes our definition of the outcome associated with each individual (aircraft MDS) that is a member of an organization (command) that generates measurements each fiscal year. In longitudinal analysis, fiscal year represents a "wave" of data. If observations are first differenced across fiscal year, the effect of aircraft mission and flyaway cost, which remain constant for each aircraft MDS, are effectively subtracted from the model. The effect of command is also removed from the model.

As a result, a first-difference model relating the change in net sales can be related to the change in the following variables: flying hours, ASD, landings per sortie, and MDS age. The results are shown in Table A.2.

For logarithmic variables, we note that the difference in logs is equivalent to the log of the ratio. One complication associated with first differencing is that the error term of the original model may not possess ideal properties. In fact, the discussion of serial correlation in the main body of the report is evidence that the error term of the original model does not have all the desirable properties. For Model 2, summarized in Table 4.3, we chose a simple, serially correlated error structure represented in Equation (4.2).

In the first-difference analysis, an "Unstructured" residual variance was used to capture the fact that the full error structure is likely more complex than can be modeled with a simple "first-order" serial correlation structure.

While the coefficient of first difference of flying hours is somewhat smaller than found in previous models, it is statistically significant. The coefficient of the first difference of ASD remains negative and significant. While the coefficient of landings per sortie first difference remains positive, it is no longer statistically significant.

The most significant finding, however, may be the fact that the aging effect is estimated to be about 2.2 percent per year and is nearly statistically significant. This first-difference approach, therefore, helps support the view that aging is a real phenomenon and not an artifact of this longitudinal analysis in which cross-sectional data are being measured over time. Any "non-stationarity" that may be present in the data is being addressed through this first-difference procedure.

Table A.2
First-Difference Model

Explanatory Variable	Coefficient	t-Statistic	Significance
$Ln(FH_t/FH_{t-1})$	0.673	7.947	0.000
$Ln(ASD_t/ASD_{t-1})$	−0.404	−2.752	0.007
$Ln(LANDSTY_t/LANDSTY_{t-1})$	0.113	0.901	0.369
$Age_t - Age_{t-1}$	0.022	1.928	0.057

NOTE: Dependent variable: $Ln(Net\ Sales_t/Net\ Sales_{t-1})$, FY04 MSD$.
Unstructured residual covariance.

References

AFTOC-Battelle, *AFTOC Logistics Distribution Table: A Plain English User's Guide*, prepared by Battelle Memorial Institute, 2004.

Freedman, David A., *Statistical Models: Theory and Practice*, New York: Cambridge University Press, 2005.

Hildebrandt, Gregory G., and Man-Bing Sze, *An Estimation of USAF Aircraft Operating and Support Cost Relations*, Santa Monica, Calif.: RAND Corporation, N-3062-ACQ, 1990.

Keating, Edward G., and Frank Camm, *How Should the U.S. Air Force Depot Maintenance Activity Group Be Funded? Insights from Expenditure and Flying Hour Data*, Santa Monica, Calif.: RAND Corporation, MR-1487, 2002. As of November 17, 2006:
http://www.rand.org/pubs/monograph_reports/MR1487/index.html

Lively, William, "What the Analyst Needs to Know and More," presented at AFTOC User's Conference, Battelle Memorial Institute, 2004.

Peltz, Eric, Lisa Colabella, Brian Williams, and Patricia M. Boren, *The Effects of Equipment Age on Mission-Critical Failure Rates: A Study of M1 Tanks*, Santa Monica, Calif.: RAND Corporation, MR-1789-A, 2004. As of November 17, 2006:
http://www.rand.org/pubs/monograph_reports/MR1789/index.html

Pyles, Raymond A., *Aging Aircraft: USAF Workload and Material Consumption Life Cycle Patterns*, Santa Monica, Calif.: RAND Corporation, MR-1641-AF, 2003. As of November 17, 2006:
http://www.rand.org/pubs/monograph_reports/MR1641/index.html

Simon, Herbert A., "Causality in Economic Models," in John Eatwell, M. Milgate, and P. Newman, eds., *The New Palgrave: Econometrics*, London: Macmillan, 1990, pp. 50–53.

U.S. Air Force, Air Force Instruction (AFI) 65-503, Logistics Cost Factors, Table A2-1, published annually.